The Prentice-Hall Science Education Series

This series reflects an ongoing commitment to science education, offering teachers, students, and the general public critical information, instruction, and activity resources that activate effective and enjoyable learning in science and mathematics. The series is designed for use by children of all ages and the adults who work with them in both school and non-school settings. These practical science handbooks give readers basic concepts, ideas, and applications to meet the challenges of the future. Among the titles:

Teaching Children About Science:
Ideas and Activities Every Teacher and Parent Can Use
Elaine Levenson

Teaching the Fun of Physics: 101 Activities
to Make Science Education Easy and Enjoyable
Janice Pratt VanCleave

The Earth and How It Works: A Lab Manual and Workbook
with Teaching Ideas, Projects, and Activities in Environmental Science
Philip R. Holzinger

The Rocket Book: A Guide to Building and Launching
Model Rockets for the Space Age
Robert L. Cannon and Michael A. Banks

THE EARTH AND HOW IT WORKS

**A Lab Manual and Workbook
with Teaching Ideas, Projects,
and Activities in Environmental Science**

PHILIP R. HOLZINGER

A Spectrum Book

Prentice-Hall, Inc., Englewood Cliffs, New Jersey 07632

Library of Congress Cataloging in Publication Data

Holzinger, Philip R.
    The earth and how it works.

    (The Prentice-Hall science education series)
    "A Spectrum Book."
    Bibliography: p.
    Includes index.
    1. Earth sciences—Laboratory manuals.
I. Title.   II. Series.
QE44.H65   1985   551'.07'8   85-3382
ISBN 0-13-223447-5

Editorial/production supervision by Marilyn E. Beckford
Cover design by Hal Siegel
Chapter-opening design by Susan Maksuta
Manufacturing buyer: Carol Bystrom

This book is available at a special discount when ordered in
bulk quantities. Contact Prentice-Hall, Inc., General
Publishing Division, Special Sales, Englewood Cliffs, N.J. 07632.

Prentice-Hall International (UK) Limited, *London*
Prentice-Hall of Australia Pty. Limited, *Sydney*
Prentice-Hall Canada Inc., *Toronto*
Prentice-Hall Hispanoamericana, S.A., *Mexico*
Prentice-Hall of India Private Limited, *New Delhi*
Prentice-Hall of Japan, Inc., *Tokyo*
Prentice-Hall of Southeast Asia Pte. Ltd., *Singapore*
Whitehall Books Limited, *Wellington, New Zealand*
Editora Prentice-Hall do Brasil Ltda., *Rio de Janeiro*

CONTENTS

PREFACE

"Spaceship Earth," as it is sometimes called, is a planet currently facing an environment crisis. Its population is rising at a rate faster than its inhabitants can deal with. Its mineral resources and energy resources are being depleted at an alarming rate, and pollution is being produced at a rate faster than the environment can handle. Because of this, the overall health of the earth can only be described as poor to fair.

To treat an ailing planet, we must first understand it. The exercises in this book are designed with that purpose in mind. Each exercise will acquaint you with some aspect of the earth and how it works. As you complete each exercise, your knowledge of our planet will expand.

When you have completed all the exercises in this workbook, you should be more aware of our environmental crises and, I hope, will want to work to improve the health of our planet. You can do this in your own life by taking steps to conserve energy, such as turning off lights when they are not in use or by lowering the thermostat in the winter. If you drive a car, you can plan your trips to conserve energy, or you can use public transportation instead. By developing an awareness of the ways in which you use energy, you will find you can save a great deal of that energy through simple conservation measures. At the same time, you will be polluting the environment less.

A second level of environmental awareness involves choosing products that are less damaging to the environment than others. A car that gets thirty-five miles per gallon, for example, is much preferable to one that gets only fifteen mpg. As another example, returnable bottles require much less energy over their lifetime than throwaway bottles or cans. Since these bottles are reusable, fewer of them need be produced, and hence fewer natural resources and energy must be used. This also means that less pollution is produced. In the area of tools and appliances, energy-requiring devices, such as snow blowers and leaf blowers, can be replaced by simpler devices, such as the snow shovel and rake. These simpler tools require less energy and resources to manufacture, and using them is generally good exercise. As another example, showerheads that use five gallons per minute or more can be replaced by ones that use only two gpm. This simple switch will save not only tremendous amounts of water but also the tremendous amounts of energy required to heat that water.

Perhaps the highest level of environmental awareness you can achieve involves adopting a world view that takes the environment into account. By adopting such a view, you can better judge the decisions made by leaders and the effect these decisions will have on the environment. You can then make your opinion heard by attending meetings, writing letters, or perhaps becoming a decision maker yourself.

Each of the exercises in this book is designed to be done in one sitting. Questions throughout each chapter test both your knowledge of the material and your common sense. A summary appears at the end of each exercise, followed by a list of activities that pertain to the subject. Answers to the questions are located in the back of the book, as is a bibliography of pertinent books.

It is not critical that the order of exercises in the book be followed closely. Each exercise is designed to be done separately, and each assumes no previous knowledge. Several of the chapters are related, however. Exercises two and three are closely related, as are exercises five and six, seven and eight, nine and ten, twelve and thirteen, and fifteen to eighteen. These should be self-evident when looking at the Contents. Exercise one, The Earth in Space, is generally a good place to start, but the order after that can be varied.

The topics in this book were chosen with the aim of educating you as to various aspects of the earth and its environment. Literally hundreds of books were reviewed, and the topics covered here were chosen from themes that were repeated over and over. It is hoped that this book will give you a solid foundation for each of these basic themes.

I would like to thank Joseph Holzinger for his help with the manuscript and for providing me with crucial materials. Mary Kennan, Christine McMorrow, and Marilyn E. Beckford are to be thanked for their editorial work, and Jim Muschlitz, Don Price, and Carl Smith are to be thanked for critically reviewing the manuscript. Thanks are also in order to Jack Crane for many of the illustrations in the book and to Fran Lawrence for several of the maps.

THE EARTH IN SPACE

At one time it was thought that the earth was the center of the universe. According to this theory, the sun, the moon, the stars, and all the planets revolved around the earth. Humankind basked in this notion, and it is easy to see why the idea was popular. Today, however, we know that the earth is not the center of everything. The earth revolves around the sun, not vice versa, and the sun is but one star in our galaxy, the Milky Way Galaxy. There are about 100 billion other stars in our galaxy. The Milky Way Galaxy, in turn, is but one galaxy in the universe. There are at least a billion other galaxies in the universe. So, as you can see, our view of the heavens has changed considerably from what it once was.

1. Our sun is just one star in the Milky Way Galaxy. About how many other "suns" are in our galaxy?

2. Until recent years, it was thought that our sun was the only star that had planets around it. But the task of finding planets revolving around

FIGURE 1-1. A view of the heavens. The three largest objects are distant galaxies. Individual stars are part of our own Milky Way Galaxy. (Palomar Observatory Photograph.)

other stars is very difficult because of the small size of planets and the large distance to stars. Recently, however, scientists have discovered what they believe to be planets revolving around a star that is relatively close to us. Does this finding increase or decrease the likelihood that there is life in other parts of the universe?

3. If scientists were to discover intelligent life on another planet, what might be one of the problems encountered in communicating with them?

The Solar System

Our solar system began forming about 5 billion years ago. It formed from a large cloud of gas and dust that began to contract. As this cloud contracted, it began to spin and in so doing flattened into a disk (Figure 1-2).

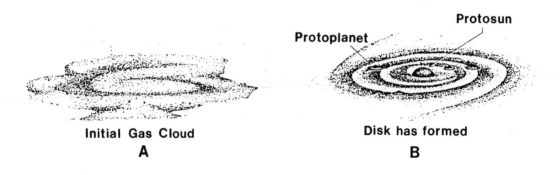

Initial Gas Cloud
A

Protosun

Protoplanet

Disk has formed
B

FIGURE 1-2. Formation of the solar system. (Pananides and Arny, 1979.)

At the center of this rotating disk, the heat was intense enough to fuse hydrogen atoms together to form helium. Once this reaction started, enormous quantities of heat and light were given off and our sun began shining.

Within the spinning disk, small planets began to form. These protoplanets grew larger as their gravity began to capture remaining dust particles. Many of the remaining gases were then driven off into space by the newly formed sun. The result was the solar system as we know it today.

4. Fusion power is a resource that we have not yet harnessed here on earth. In a fusion reactor, hydrogen atoms are forced together to make helium atoms. Although great amounts of heat are needed for this reaction to take place, once it does, tremendous quantities of energy are released. Where in our solar system is fusion currently taking place?

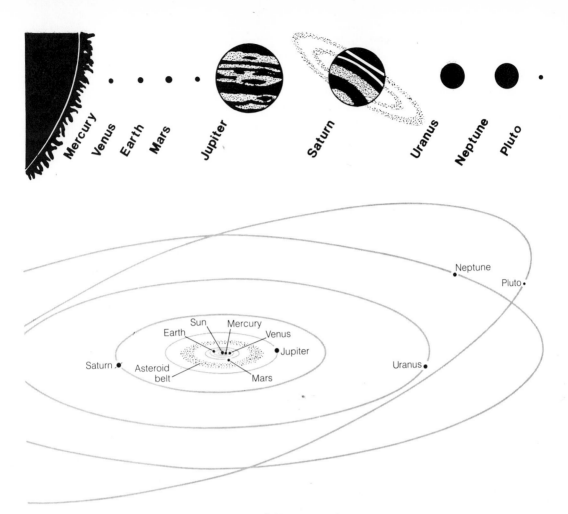

FIGURE 1–3. Our solar system. (Press and Siever, 1974.)

5. Of the nine planets in our solar system, four of them are many times larger than the rest. These four are referred to as the giant planets. What are their names?

6. There is an area in our solar system consisting of thousands of boulders that are orbiting the sun. These boulders either never formed into a planet or are the remnants of a planet that broke apart. What is the name of this zone of boulders?

7. Our moon and the planet Mercury have a lot in common. Both are about the same size, and both are extensively cratered. This is because neither has an atmosphere to burn up incoming meteors. Mercury does differ from the moon in that its surface is much hotter. Why is this?

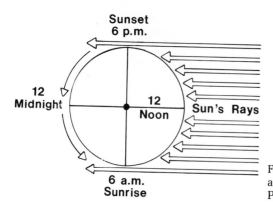

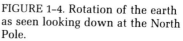

FIGURE 1–4. Rotation of the earth as seen looking down at the North Pole.

The Earth

The earth *rotates* in a counterclockwise direction as shown in Figure 1–4.

Notice in Figure 1–4 that the time of day for a particular area depends on where it is in relation to the sun. For example, when the sun is directly overhead, it is noon.

8. When it is 3 P.M. in Phoenix, Arizona, what time is it in Bombay, India, halfway around the world?

In addition to rotating, the earth also *revolves* around the sun (Figure 1–5).

FIGURE 1–5. The earth revolving around the sun.

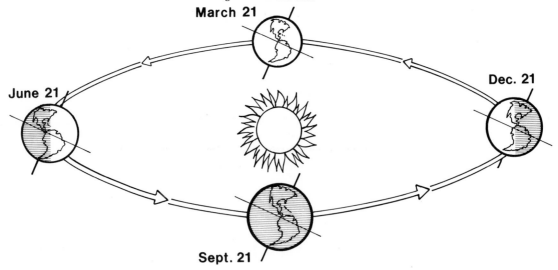

4

9. How long does it take for the earth to rotate once on its axis? How long does it take the earth to revolve once around the sun?

The North–South axis upon which the earth rotates is not perpendicular to the sun. Instead, it is inclined at an angle of 23½° (Figure 1–6).

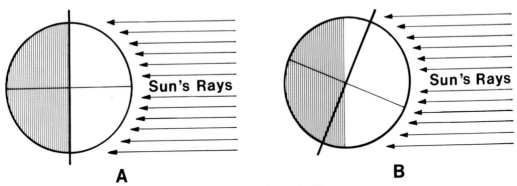

A   **B**

FIGURE 1–6. The earth's axis is tilted at 23½° as shown in (B).

The fact that the earth's axis is tilted accounts for the seasons we experience. Notice in Figure 1–5 that on December 21 the sun is shining mainly on the southern hemisphere and on June 21 it is shining mainly on the northern hemisphere. Rotation, of course, still accounts for day-time and night-time in each hemisphere.

10. As the earth moves from its position on March 21 to its position on June 21 (Figure 1–5), does the northern hemisphere receive more or less sunlight?

11. When it is summer in Chicago, it is winter in Cape Town, South Africa. Why?

12. When it is autumn in Chicago, what season of the year is it in Cape Town?

13. If you look closely at Figure 1–5 and imagine the earth rotating, you will see that on June 21 the North Pole and nearby areas receive sunlight twenty-four hours a day. This is true for a period of time before and after June 21 also.
     Spitsbergen is an island north of Norway near the North Pole. This island is part of the area known as the land of the midnight sun. How do you think it got that name?

14. How many hours of sunlight does Spitsbergen receive on December 21?

Summary

Our galaxy, the Milky Way Galaxy, is one galaxy in a universe that contains many galaxies. Within the Milky Way Galaxy are billions of stars, one of which is our sun. The earth is one of nine planets that revolve around the sun. As the earth rotates on its axis, we experience night and day; as the earth revolves around the sun, the tilt of its axis results in our seasons.

Latitude and Longitude

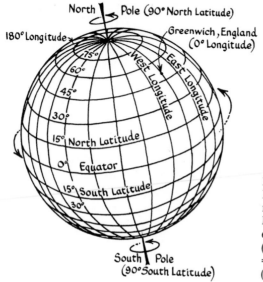

FIGURE 1-7. Latitude and longitude. Latitude is measured in degrees north and south of the equator. Longitude is measured in degrees east and west of the prime meridian (0° longitude). Each line of latitude = 15°. Each line of longitude = 20°. (Fagan, 1965.)

A point anywhere on the surface of the earth can be located by giving first its latitude, then its longitude. For example, point A is located at 40°N latitude and 50°W longitude. Its coordinates can be abbreviated as 40°N, 50°W.

15. Give the latitude and longitude coordinates for points B and C.

Activities

- Observe the night sky with a telescope.
- Demonstrate the seasons by walking a globe around the outskirts of a room. Pretend the sun is in the center of the room.
- Write a report on one of the planets.
- Sketch a picture of a spacecraft you could design.
- Make a list of items you would include inside a spacecraft for people to survive a trip to Mars.

2
THE INTERIOR OF THE EARTH

Figure 2-1 shows what the interior of the earth looks like.

> *Crust.* The crust of the earth is solid. Its composition varies.
> *Mantle.* The mantle is solid and is made up mainly of the rock peridotite.
> *Outer Core.* The outer core is liquid and is composed mainly of iron.
> *Inner Core.* The inner core is solid and is made up mainly of iron.

1. If the crust of the earth is about ten miles thick, what is the total diameter of the earth as shown in Figure 2-1?

No one has seen the interior of the earth, yet scientists have a fairly good idea of its composition. They base their evidence on three lines of thinking:

> direct observations of mantle material
> earthquake studies
> gravity studies

Let's examine each of these three lines of evidence.

Direct Observations of Mantle Material

Kimberlite Pipes. In a few places in the earth's crust, the underlying mantle has intruded. One such place is in kimberlite pipes such as those found in South Africa.

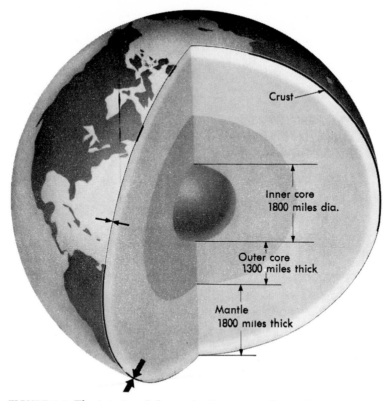

FIGURE 2-1. The interior of the earth. (Ramsey and Burckley, 1961.)

      Kimberlite pipes are made up principally of the rock peridotite, a green-to-brown glassy rock. With this rock, diamonds occasionally may be found. Diamonds form only under extremely high pressures. Such pressures occur only beneath the crust of the earth in the mantle. Because diamonds are found imbedded in the rock peridotite, it is believed that the mantle of the earth is made up of peridotite.

FIGURE 2-2. Origin of a kimberlite pipe.

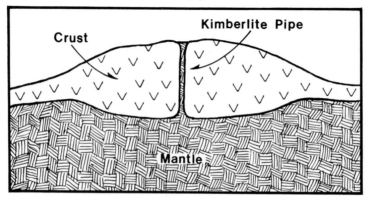

2. Why does pressure increase as you go deeper into the earth?

3. Where in the earth would you expect the greatest pressure to be?

Volcanoes. Volcanoes also yield evidence as to the composition of the mantle because they pour out lava that, in many cases, has originated in the mantle of the earth. This lava is composed of basalt, a dark rock that underlies most of the world's oceans.

You might expect that if the mantle of the earth were composed of the rock peridotite, volcanoes should erupt lava composed of peridotite. However, laboratory experiments have shown that when peridotite is partially melted, the first liquid that appears is made up of basalt. It is probable that this basaltic lava then makes its way to the surface of the earth before the peridotite has had a chance to melt completely. Thus, the fact that most oceanic volcanoes erupt basaltic lava tends to confirm that the mantle of the earth is made up of the rock peridotite.

4. The concept of only partially melting a solid can be demonstrated by putting a water-soaked sponge in the freezer. After a few hours you would have a solid "rock" that would be composed of sponge plus ice. If you then put this "rock" on the table at room temperature, it would partially melt. What would be the first liquid given off when this solid starts to melt?

5. When peridotite starts to melt, what is the composition of the first liquid given off?

Earthquake Studies

An earthquake occurs when two pieces of the earth's crust move past each other (Figure 2–3).

FIGURE 2–3. An earthquake and the shock waves it generates.

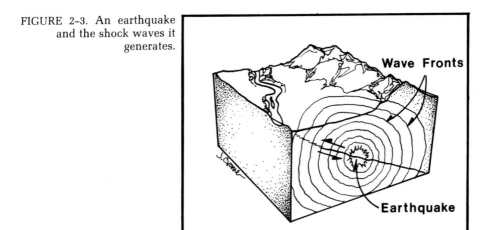

When an earthquake occurs, it sends shock waves through the entire earth. These shock waves can be felt by people only a few tens or hundreds of miles from where the earthquake occurred. However, delicate instruments can record shock waves all around the world.

Two types of shock waves, *P-waves* and *S-waves* (P for primary, S for secondary) are used in determining what the interior of the earth looks like. These two waves are shown in Figure 2–4.

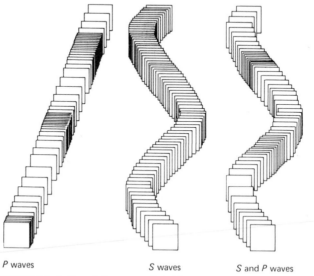

P waves   S waves   S and P waves

FIGURE 2–4. P and S waves. (McAlester, 1973.)

You can demonstrate P and S waves for yourself by stretching out a Slinky. Pushing on the Slinky will create a P wave, while shaking the Slinky back and forth will create an S wave.

After an earthquake, P and S waves travel through the earth and are picked up by recorders on the surface of the earth. These recorders plot the arrival of both types of wave. As can be seen in Figure 2–5, there are certain areas of the earth where only P waves are recorded. These are called the *shadow zone.* In the shadow zone, no direct S waves are recorded.

It is known from laboratory experiments that S waves do not pass through liquids. Thus, the presence of a shadow zone for S waves signals to scientists that the outer core of the earth is liquid. It is this liquid that prevents the passage of S waves.

6. The inner core of the earth is thought to be solid because P waves travel abnormally fast through it. It is known that S waves travel through solids, yet S waves do not travel through the inner core of the earth. Why not?

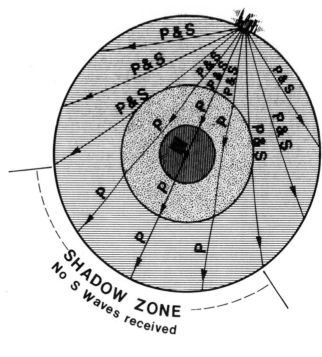

FIGURE 2-5. The passage of P and
S waves through the earth following
an earthquake.

It has been noted that earthquake waves change markedly in speed at
a depth of about ten miles. Because it is known that these waves travel at different speeds through different materials, it is believed that this ten-mile depth
is the boundary between the crust and the mantle of the earth.

Ten miles is the average thickness of the crust. The crust can vary in
thickness, however, from seven miles thick under the oceans to about twenty
miles thick under the continents (Figure 2-6).

FIGURE 2-6. Variations in thickness in the earth's crust.
(Strahler, 1970.)

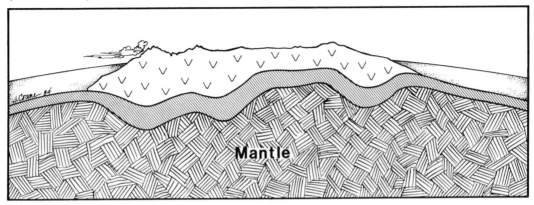

7. In the 1960s there was a project in which scientists were to drill through the crust of the earth into the mantle. Although this project was never actually attempted, why do you think these scientists were planning to drill offshore instead of on land?

The mantle of the earth is a solid that is more rigid than steel. Yet, near the top of the mantle (but still within the mantle) is a layer that is solid but behaves like a liquid in some ways. This layer, called the zone of partial melting, was discovered when scientists found that some P and S waves from earthquakes were being reflected off it at depths of 60–200 miles.

The zone of partial melting behaves in some ways like tar. Tar is a solid, yet it flows when a weight is put on it for a long time. Likewise, the zone of partial melting is a solid that flows very slowly when weight is applied. In this case the weight is that of the overlying crust.

The next exercise details how the continents are slowly drifting. Keep in mind when reading the exercise that the crust and upper mantle of the earth are in effect "floating" over the zone of partial melting.

8. Scientists have discovered that S waves are weakened when they pass through the zone of partial melting but that P waves are not. Why should S waves be weakened but not P waves?

Gravity Studies

A third way of determining the composition of the earth is by gravity studies. It is known that all objects exert a pull of gravity, however slight it may be. It is also known that the denser an object is, the greater its pull of gravity. Thus, a steel ball exerts a greater pull of gravity than does a wood ball of similar size, although the pull of gravity on either is difficult to measure because of their small size.

When dealing with an object the size of the earth, the makeup of its interior will be reflected in how it pulls nearby objects such as the moon, asteroids, and satellites.

FIGURE 2-7. How density affects gravitational pull: (A) A planet of low density alters the path of a stray asteriod only slightly; (B) A dense planet has a greater pull on an asteroid passing by; (C) A planet of high density actually "captures" a passing asteroid.

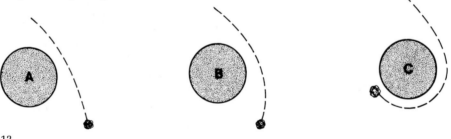

The earth's pull on nearby objects has been measured, and from these measurements it has been determined that the earth, as a whole, has an average density of 5.5 grams per cubic centimeter. The earth's thin crust is known to have an average density of only 2.8 gm/cm^3, while peridotite, the rock comprising the mantle, has an average density of only 4.4 gm/cm^3. This means that if the earth as a whole has a density of 5.5 gm/cm^3, the inner and outer cores of the earth, given their size, must be composed of material that is quite dense. This material must have a density of 10–11 gm/cm^3 to average out the relatively lighter crust and mantle. The only common material known that has this required density is iron. Pure iron itself would make the earth a little too dense, but small amounts of impurities, such as nickel, would give it the required density. It is from these gravity studies that scientists have come to believe that the inner core and outer core of the earth are comprised chiefly of iron.

9. Assuming the inner core and outer core of the earth to be made of iron, would you weigh less or more than you do now if the earth's mantle were also made of iron instead of peridotite? Why?

Summary

The interior of the earth has been deciphered using three lines of evidence—direct observations of mantle material, earthquake studies, and gravity studies. Direct observations of mantle material include those rocks brought to the surface from volcanoes and kimberlite pipes. Earthquake studies concentrate on the behavior of P and S waves through solids and liquids. Gravity studies deal with the earth's gravitational attraction of objects near to it.

10. Temperatures within the earth are believed to range from 1200° C in the zone of partial melting to 4000–4500° C in the inner core. These temperatures would melt rock at the earth's surface. But at great depths, pressure is a factor that acts to keep the rock solid. Why is it difficult to determine the temperature at the center of the earth?

Activities
- Put a weight on a mass of Silly Putty. Observe what happens at five-minute intervals. This Silly Putty has somewhat the same consistency as the zone of partial melting.
- Demonstrate P and S waves with a Slinky.
- Make or draw a model of the earth.
- Examine a piece of peridotite. What minerals make up this rock?

3
CONTINENTAL DRIFT

On a world map, South America and Africa look as though they could fit together like a jigsaw puzzle. Alfred Wegener, a German meteorologist, was the first person to suggest that these two continents were indeed at one time together, then later broke apart and drifted to their present locations. He based his theory of continental drift on the similarity in *fit* between the coastlines of the two continents and identical *fossils* found on both these continents. Wegener reasoned that the fossil similarity could be accounted for only if the two continents had once been connected.

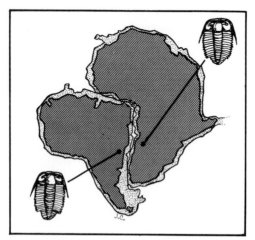

Figure 3–1.
Alfred Wegener
and his theory.

1. What is a fossil?

Wegener proposed his theory of continental drift in 1912. During his lifetime the theory was not accepted, and Wegener was severely criticized for proposing it. In the late 1950s, however, new information was collected that supported Wegener's idea that the continents may be drifting. In this chapter you will work with some of this information.

Convection Currents

Figure 3-2 shows the probable way that South America split from Africa. Convection currents in the mantle of the earth were responsible for the split. These currents are similar to the currents you would generate if you put a Bunsen burner under a beaker of water.

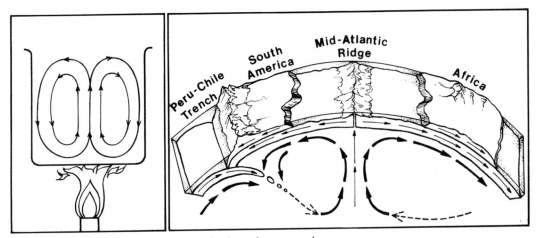

FIGURE 3-2. Convection currents in a beaker of water and in the earth.

Notice in Figure 3-2 that a crustal plate can be made up of both ocean and continent. Where two plates collide, such as on the left side of South America, one plate is forced beneath the other. This is called a *subduction zone*. As the down-turned plate returns to the mantle of the earth, part of it may melt and send hot lava to the surface. The Andes mountains of South America are the result of a collision between two crustal plates.

2. In Figure 3-2, what is the name given to the feature where the two crustal plates are spreading away from each other?

3. What is the name given to the feature just west of South America where the one plate is being forced beneath the other?

Magnetic Stripes

As molten rock cools, it begins to harden into solid rock. At a temperature of about 500° C, while the rock is still partially molten, magnetic minerals in the rock align themselves with the magnetic field of the earth—just as a compass does.

At times in the past millions of years, the magnetic field of the earth has been reversed from what it is today. During such times, a compass would have pointed south instead of north. Nobody knows exactly what caused the magnetic field to reverse; all we know is that it occurred.

Evidence for this magnetic field reversal is found in the ocean floor rocks. These rocks show a magnetic striping pattern that could have resulted only if the rocks formed from a spreading ridge as shown in Figure 3–3.

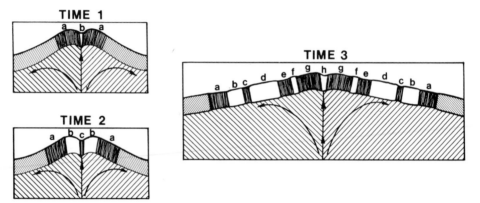

FIGURE 3–3. The formation of magnetic stripes. By Time 3, the magnetic field has reversed itself seven times. Four episodes of normal polarity (black) can be seen, as well as four episodes of reversed polarity (white).

It was this magnetic striping pattern in the ocean rocks that convinced many scientists that continental drift was taking place.

4. How many periods of normal polarity are shown in Figure 3–4, including the present period?

5. How many periods of reversed polarity are shown?

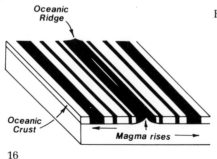

FIGURE 3–4. (Flint and Skinner, 1977.)

Crustal Plates

The locations of the major crustal plates on the earth are shown in Figure 3–5, along with the direction each is moving. Many features on the earth can be explained in terms of these plates.

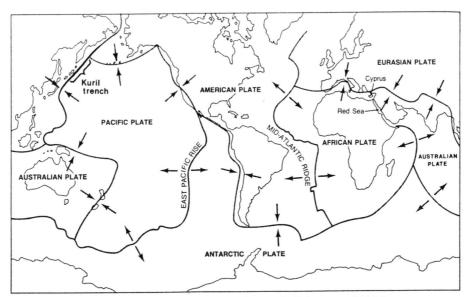

FIGURE 3–5. The major crustal plates. (Ehrlich, P., Ehrlich, A., and Holdren, 1977.)

6. The San Andreas Fault in California is an area where one crustal plate is slipping past another. The San Andreas Fault lies on the boundary between what two major plates?

FIGURE 3–6. Subduction at the Kuril trench.

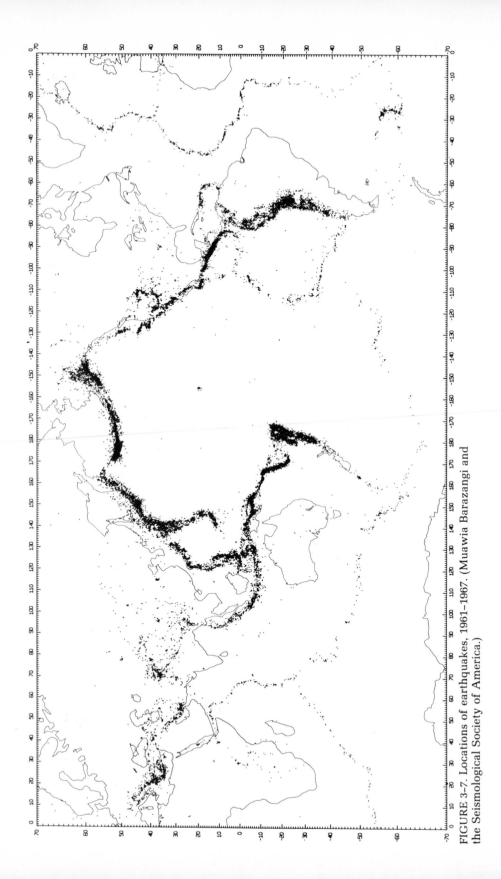

FIGURE 3–7. Locations of earthquakes, 1961–1967. (Muawia Barazangi and the Seismological Society of America.)

7. The Kuril trench in the Pacific Ocean runs from northern Japan to the Kamchatka peninsula of Russia (Figure 3-5). This trench has formed because part of the Pacific plate is being subducted beneath the Eurasian plate (Figure 3-6). Behind the Kuril trench are the Kuril Islands. Notice in Figure 3-5 that these islands run parallel to the trench. How did these islands form?

8. The Himalayas in Asia are the highest mountains in the world. They are the result of the collision of what two drifting plates?

Earthquakes

Where there is a junction between two plates, a great deal of friction builds up. This is true whether the plates are spreading, colliding, or simply sliding past each other. When two moving plates overcome the friction between them, an earthquake occurs.

9. Figure 3-7 shows the locations of 30,000 earthquakes that occurred between 1961 and 1967. Each dot on the map shows the location of an earthquake. Compare the location of these earthquakes with the plate boundaries in Figure 3-5. What comparison can you make?

Summary

The theory of continental drift was introduced by Alfred Wegener in 1912. This theory states that convection currents in the mantle of the earth cause plates of the earth's crust to slowly drift. The location of earthquakes and the pattern of magnetic stripes on the ocean floor both serve as strong evidence for continental drift. Many topographic features on the earth's surface can be explained by this theory.

10. Spreading rates for the drifting continents vary from about 1 cm/yr–10 cm/yr. South America and Africa are spreading from each other at a rate of about 2.5 cm/yr. Given that the distance between them today is about 5,000 km, calculate the time it took these two continents to drift to their present locations, assuming they were once together. (Conversion Factor: 1km = 100,000 cm.)

Activities
- Trace the flow of water in a beaker as it is heated. Note the convection currents.
- Research other evidence pertaining to the theory of continental drift.
- Examine a topographic map of the world. Make comparisons between the topography and Figure 3-5.
- Read about the geologic history of the Hawaiian Islands. A good place to begin your reading would be Frederic Martini's *Exploring Tropical Isles and Seas* (Prentice-Hall, Inc., publishers).

4

THE OCEANS

The oceans cover 71 percent of the earth. The area they cover is not a feature-less plain as was once thought. Instead, it contains topography that is in places more rugged than that found on land. As shown in Figure 4–1, the ocean floor contains large oceanic ridge systems, numerous volcanoes, and in places deep trenches.

FIGURE 4–1. The ocean floor. (Flint and Skinner, 1977.)

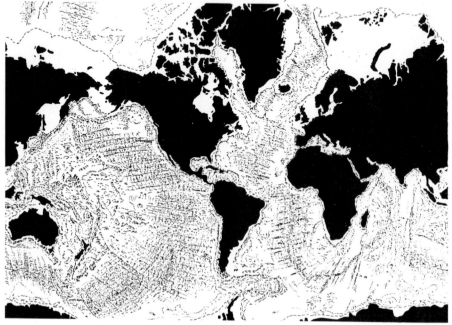

Many of the features of the ocean floor have been formed by the movement of the earth's crustal plates. Where plates have been moving apart there are mountain ranges, such as the Mid-Atlantic Ridge and the East Pacific Rise southwest of South America. Crustal spreading taking place over millions of years accounts for the great width of some of these underwater mountain systems. Where plates have been converging on each other, and one plate is being subducted (Figure 3–6), there are deep trenches such as the Peru-Chile trench off the coast of South America and the Kuril trench northeast of Japan.

1. The picture of the ocean floor shown in Figure 4–1 was obtained by taking countless depth measurements from ships. Since the 1920s these measurements have been made by using *echo sounders*. An echo sounder operates by bouncing sound waves off the ocean floor. The length of time for the sound wave to return is noted, and the depth of the ocean floor can then be calculated.

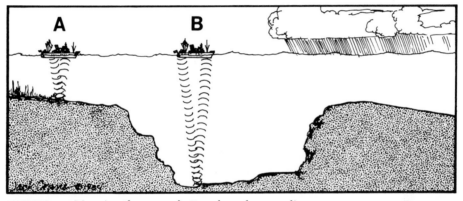

FIGURE 4–2. Mapping the ocean bottom by echo sounding.

At location A, the sound wave took 4 seconds to return to the ship, indicating that the depth to the ocean bottom was 10,000 feet. At location B the sound wave took 6½ seconds to return to the ship. What was the depth to the ocean bottom at location B?

Continental Shelves

Continental shelves are gently sloping areas found adjacent to continents. These shelves are in relatively shallow water and are made up of material that was eroded from the continents.

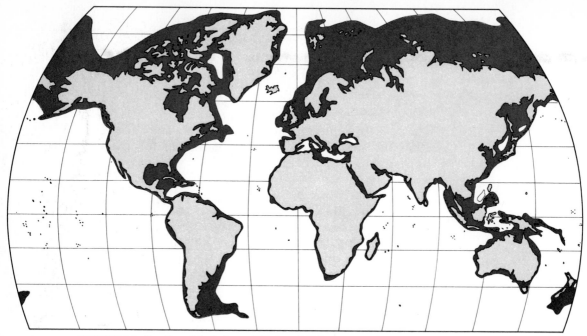

FIGURE 4-3. Continental shelves. (Flint and Skinner, 1977.)

During past ice ages, more of the earth's water was found on the continents in the form of ice. As a result, sea level was lower and large areas of the continental shelves were exposed as land.

2. Life is not evenly distributed throughout the oceans. Rather, 90 percent of all ocean life is found on the continental shelves. Ironically, most of the pollution also takes place here. Name one way that people pollute the waters of the continental shelf.

Chemistry of Sea Water

Sea water is composed of 96.5 percent water and 3.5 percent dissolved elements (Figure 4-4).

3. As shown in Figure 4-4, the major elements dissolved in sea water are sodium and chloride. These two elements are responsible for the characteristic taste of sea water. They combine together to form what common product?

4. Among those elements listed as "Others" in Figure 4-4 are rare elements such as gold. It is estimated that there are millions of tons of gold dissolved in sea water. Removing this element from the other impurities in the water has proved to be nearly impossible, however.

    If all the gold dissolved in the oceans could be removed, gold would become a worthless commodity. Why?

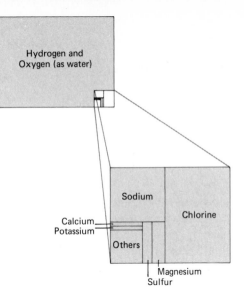

FIGURE 4-4. The composition of
sea water. (McAlester, 1973.)

Currents

Most of the water in the oceans is in constant motion. At the surface, this motion is determined largely by global *wind patterns* and the *position of the continents* (Figure 4–5).

FIGURE 4–5. Major surface currents. (U.S. Navy Hydrographic Office.)

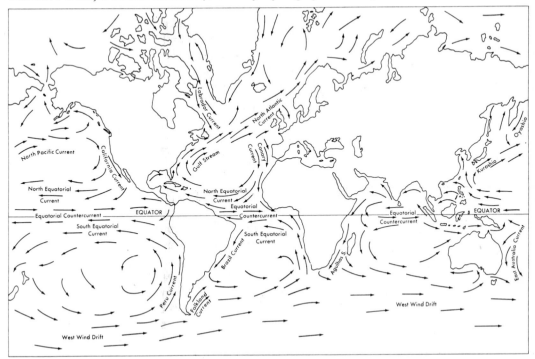

The surface currents of the oceans, along with the atmosphere, are responsible for the transfer of heat from the equator toward the poles of the earth. Without this heat transfer, tropical areas would be much hotter than they are and polar areas would be much colder.

5.   Warm currents are those that flow from the equator toward the poles. The water in these currents is cooled considerably before it returns toward the equator as a cold current. Name two cold currents off the coast of South America.

In addition to surface currents, there are deep currents in the ocean. Many of these currents originate in polar areas where the water becomes quite cold and dense. As a result it sinks toward the bottom and flows back toward the equator as a bottom current. Other currents arise because of salinity differences. This is the case in the Mediterranean Sea, for example, where high evaporation and low rainfall cause the water in that area to have a high salinity and a high density. This water is heavier in weight than water in the Atlantic Ocean and therefore flows into the Atlantic as a bottom current (Figure 4-6).

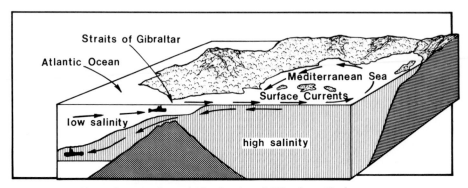

FIGURE 4-6. Flow of water through the Straits of Gibraltar. (Stokes and Judson, 1968.)

Water leaving the Mediterranean Sea through the Straits of Gibraltar is replaced by less dense water from the Atlantic Ocean. This Atlantic water flows as a surface current into the Mediterranean Sea. Thus, two opposite flowing currents are present in the Straits of Gibraltar.

6.   In World War II, Allied ships were stationed in the Straits of Gibraltar to listen for the engines of German submarines that may have been trying to enter the Mediterranean Sea. Some German subs were able to pass both ways through the straits, however, without being detected. Why were the Allied ships unable to hear the sounds of their engines?

24

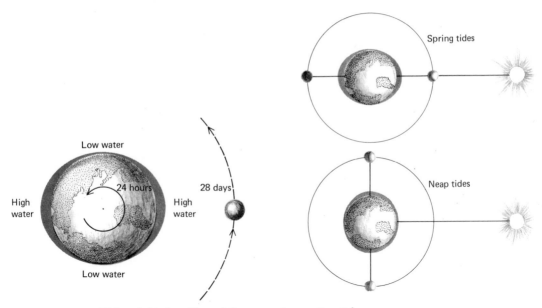

FIGURE 4-7. Tides: (left) the effect of the moon in creating tides; (right) the effect of the sun and moon in creating tides. (McAlester, 1973.)

Tides

Tides are a result of the *gravitational attraction* of both the moon and sun on the earth. The moon's attraction is greater than the sun's because the moon is much closer to us. As shown in Figure 4-7, the moon creates a bulge in the ocean waters on the side that faces it. It also tends to pull the earth "out from under" the opposite side, thereby creating a bulge there also.

When the sun and moon are pulling on the earth in the same line, the result is higher than normal tides. These are called spring tides. (*Note:* Spring tides have nothing to do with the season of the year. That is just the name that was given them.) When the sun and moon are pulling at right angles to each other, we have lower than normal tides called neap tides.

In Figure 4-7 the bulge of the ocean waters is greatly exaggerated. In reality, the oceans bulge out less than three feet. As the earth rotates, however, this tidal bulge can become "funneled" into river inlets. If the shape of the inlet is right, the height of the tide can be greatly magnified. This is the case in the Bay of Fundy in Nova Scotia where the tidal range is as great as fifty feet (Figure 4-8).

7. In the Bay of Fundy, why would it be important for fishermen to know when the tide was going to be low?

FIGURE 4–8. The Bay of Fundy: (top) high tide; (bottom) low tide. (Flint and Skinner, 1977.)

Waves

Waves are generated by the wind. The strength and duration of the wind affects the size of the waves as does the area over which these winds blow. Larger waves can be generated in the Pacific Ocean, for example, than in the Great Lakes because the wind has a much greater area over which to work.

When waves reach the shoreline they erode and sculpture it. Many factors are involved in how the shoreline will look, thus a whole branch of science is devoted to the study of shorelines.

8. In Figure 4–9, shorelines A and B look alike in many respects. Yet shoreline B is much less susceptible to abrasion by waves than is shoreline A. Why?

A   B

FIGURE 4–9. Two shorelines.

Life in the Oceans

Microscopic plants called *phytoplankton* serve as the base for the food chain in the ocean. These plants serve as food for small animals called *zooplankton*. Both phytoplankton and zooplankton are the food supply for much larger animals.

      Although Figure 4–10 shows many different life forms, it should be remembered that phytoplankton make up 90 percent of the weight—called *biomass*—of all life in the oceans.

FIGURE 4–10. Ocean life.

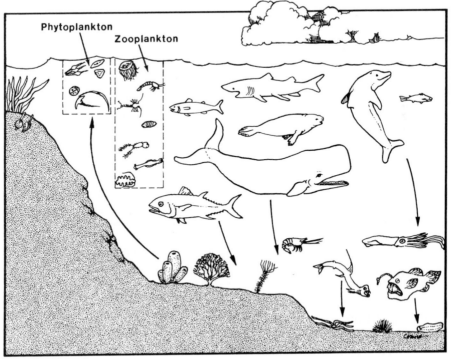

9. Organisms that live at the surface eventually die and their remains sink toward the ocean bottom. Thus, the lower ocean layers receive a constant "rain" of nutrients from above. In certain areas of the world, such as off the coast of Peru, these lower ocean layers are forced to the surface by currents. The rich waters then yield a bloom of phytoplankton that, in turn, allows other organisms such as fish to multiply. This upwelling of waters is very important to the fishing industry.

In the future, how might pumping water from the depths of the ocean to the surface allow us to "farm" the sea?

Summary

The oceans cover 71 percent of the globe. The area they cover is marked by topography that is at least as rugged as that on land. Continental shelves are gently sloping areas adjacent to continents. Ninety percent of the life in the oceans is found on these shelves. Ocean water is made up of 96.5 percent water and 3.5 percent dissolved elements. This water is in constant motion. Winds direct surface currents while differences in water density drive currents beneath the surface. The tides are caused by the gravitational pull of the moon and sun. Waves are caused by the wind and they act to "sculpture" a shoreline. Life in the oceans is quite varied and is supported directly or indirectly by microscopic plants called phytoplankton.

10. Oil tankers pose a pollution threat to ocean waters if they are involved in an accident. When this happens, large quantities of oil may spill into the water and cause great havoc to the ecosystem. For example, such an accident may wreck both the fish and shellfish industry in an area.

The trend today is toward building larger tankers that can transport more oil at one time. Why might it be a good idea to stick with the smaller tankers instead?

Activities

- On January 23, 1960 two men descended to the bottom of the Marianas trench, the deepest spot in the world's oceans. Their steel balloon-like vessel, called a bathyscaph, rested on the bottom at a depth of 35,800 feet. This is almost seven miles deep! If the weight of overlying water causes a pressure increase of one atmosphere for every thirty-three feet of depth, how many atmospheres of pressure were exerted on the sides of this bathyscaph at this depth?
- Compare the ocean currents (Figure 4–5) with the global surface wind patterns (Figure 6–6).
- Rachel Carson, in her book *The Sea Around Us* described the oceans as "the earth's thermostat." What did she mean by this?
- "Create" sea water by adding 3.5 grams of salt to 96.5 grams of pure water. Then taste it.

5
THE HYDROLOGIC CYCLE

If you wash a blackboard, then observe it as it dries, you realize that *evaporation* is taking place. Evaporation causes the water to leave the blackboard and enter the air. You may think of it as the blackboard becoming drier and the air around it becoming a little wetter or more humid.

Evaporation from the earth's surface occurs just as it does from the blackboard. Over 70 percent of the earth's surface is covered by water, so you can see that quite a lot of water vapor can enter the earth's atmosphere at any one time. This water vapor begins the hydrologic cycle, the cycle through which water moves on the earth.

FIGURE 5-1. The hydrologic cycle. (Ahrens, 1982.)

Condensation

Precipitation   Wind ←

Transpiration

Evaporation

Glaciers

Lakes

Groundwater

Runoff

Oceans

Study closely in Figure 5-1 how water moves in the hydrologic cycle. Water enters the atmosphere by evaporation from the oceans, rivers, and lakes. It also enters the atmosphere by transpiration through the leaves of plants. Indeed, trees and other plants transpire enormous quantities of water to the atmosphere, especially during the summer months.

1. In some climates, why do many outdoor plants transpire enormous quantities of water in the summer and almost none in the winter?

Once water is in the atmosphere, it can then form into clouds. When conditions are right in a cloud, it gives up its water in the form of rain or snow, depending on the temperature. Under rarer conditions, sleet or hail may be the form of precipitation.

Once precipitation lands on the ground, it begins its journey back to the sea. As you can see in Figure 5-1, some of this precipitation never returns to the sea, because it either evaporates back to the atmosphere directly or is taken up by plants and transpired back to the atmosphere in that way. Moisture that is not evaporated or transpired will return to the sea, however. It returns in one of three ways—streams, glaciers, or ground water.

2. When it rains over the ocean, the hydrologic cycle may be thought of as being quite simple. It consists only of evaporation and what else?

3. In some areas of the world, rainfall occasionally evaporates before it ever reaches the ground. Thus, a rainstorm can be seen but not felt. Would this be more likely to happen where the surrounding air is humid or where it is dry?

Streams

Rain falling on the land can flow downhill until it reaches a stream. The stream, it reaches eventually flows into a larger stream that, in turn, joins a still larger stream, and in this way the water eventually finds its way back to the ocean. Of course, there are short cuts in this route of water flow. Some precipitation falls directly into a stream and does not run off the land first. Also, some water from the stream evaporates directly back to the atmosphere and never reaches the ocean. In addition, in the winter months, many areas receive snow. This snow may remain on the ground for long periods of time before melting. Once it melts, that portion which does not seep into the ground will run off the land into a stream.

4. Many towns and cities rely on stream or river water for their water supply. The water is withdrawn, used, then returned to its source. In industrialized countries, the average person uses about 100 gallons of water a day. Name five ways that you use water.

Glaciers

The left side of Figure 5–1 shows the hydrologic cycle in *glaciated areas*. Glaciated areas are areas where the temperature year round is so cold that snow does not melt. Such areas occur near the north and south poles, and in high mountainous areas such as the Rocky Mountains and Swiss Alps.

As one snowfall after another is deposited in these areas, a thick layer of snow develops. Due to the pressure from the overlying snowfalls, the bottom of this snow layer begins to turn to ice. As more snowfalls accumulate, more ice is formed at the bottom of the snow layer.

Once a thick layer of ice has developed in an area, this ice begins to slowly flow downhill under the force of gravity. You may not think of ice as being able to flow, but it does. This ice may be several hundred feet thick or thicker, and is, therefore, under tremendous pressure. Under such great pressure, the ice recrystallizes as it flows. It remains in its solid state, yet still flows. Its flow is not perceptible to the eye because it is usually on the order of only a few inches a day. Yet it does flow, and in this way snow that fell near the top of the mountain decades or centuries ago eventually finds its way back to the ocean.

5. As recently as 15,000–20,000 years ago, ice age glaciers covered much more of the earth's surface than they presently occupy. In the hydrologic cycle, more water was tied up as ice than is today. Would sea level have been higher or lower at that time? Why?

6. If the polar ice caps should start to melt today, why would people living in coastal areas have cause for concern?

Ground Water

Water from rain or melting snow either evaporates, runs off the land to a stream, or soaks into the ground. If it soaks into the ground it may be taken up by plant roots and transpired back to the atmosphere. If this does not happen, the water will soak further into the ground and will become *ground water*.

Ground water is water that is found in the pores and crevices in soil and bedrock. Under the pull of gravity, this water moves toward the low spots in the landscape, namely, the streams, or in coastal areas, directly back toward the ocean. This movement is usually very slow, often on the order of only a few feet a day. The ground water system will be discussed further in a later chapter.

7. Some rain falling on the land surface eventually becomes ground water. Much of it does not, however. What becomes of the rain that does not make its way into the ground water system?

8. Many coastal resort cities experience water shortages during the summer tourist season when the population may be five or ten times what it is during the off season. These communities commonly utilize ground water from wells to satisfy their water needs. Why don't they simply pump water from the nearby ocean?

To better understand the hydrologic cycle, let us examine how much water is stored in each of its parts (Table 5-1).

TABLE 5-1   Distribution of Water on the Earth.

SOURCE	PERCENTAGE OF TOTAL WATER ON EARTH
Streams	0.0001
Lakes and inland seas	0.017
Ground water	0.615
Water in the atmosphere	0.001
Ice	2.15
Water in the oceans	97.2
Total	100%

Total amount of water on earth: 326 million cubic miles

(Source: U.S. Geological Survey)

9. Rank each of the sources of water shown on Table 5-1 from the source containing the most water to the source containing the least amount of water. Beside each source list its percentage.

Summary

The hydrologic cycle describes the movement of water on the earth. The cycle begins with evaporation of water to the atmosphere, either from oceans, rivers, and lakes, or by transpiration from plants. (Most of this water is from the oceans, however.) Once in the atmosphere, this water forms into clouds, which may then drift over land. If conditions in a cloud are right, it gives up its moisture in a form of precipitation—rain, snow, sleet, or hail. After precipitation has landed on the ground, it returns to the sea by streams, glaciers, or ground water. Of course this water is subject to evaporation or transpiration anywhere along its return journey.

Most of the water on the earth—97 percent—is stored in the oceans.

10. The greatest water purification system on earth occurs when water evaporates from the oceans. Moisture leaving the oceans to form rain clouds differs from ocean water in one key respect. What is it?

Activities

- Obtain some Silly Putty and observe over the course of a class period how slowly it flows. This Silly Putty resembles how a glacier flows.
- Pour a thin layer of salt water onto a surface. Notice what remains after the water has evaporated.
- Pour a measured quantity of water onto a plant that has just begun to wilt. Note how long it takes before this plant again begins to wilt. This will let you see how fast this plant transpires water to the atmosphere.
- Take a field trip to a local reservoir and examine the hydrologic cycle first-hand.

6
THE GROUND WATER SYSTEM

Water stored in the ground is referred to as *ground water*. Ground water is used by many towns and cities for their water supply. It is also used by people who live in rural areas that are not served by a municipal water supply.

Notice in Figure 6–1 that ground water originates from water percolating down through the soil. The top surface of ground water is referred to as the *water table*. The water table may also be thought of as the level to which the ground is saturated with water. As shown in Figure 6–1, the water table corresponds to the water level in a beaker filled with sand. Above the water table,

FIGURE 6–1. Ground water.

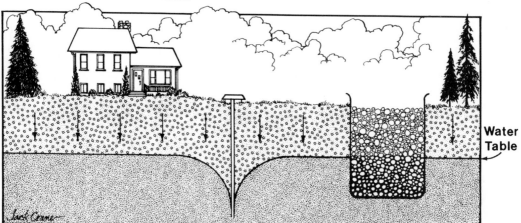

only a few of the spaces between sand grains contain water; below the water table, all of the pores between sand grains are filled with water.

Notice in Figure 6-1 that the water table has been depressed in the immediate vicinity of the well. This depression of the water table takes the shape of a cone and is thus called the "cone of depression." The size of the cone of depression varies depending on how much the well is pumped. Because of gravity, water tends to flow into the cone of depression.

1. Ground water usually moves very slowly, on the order of a few feet per day. This is why ground water pollution is such a serious problem. Polluted ground water does not readily cleanse itself. If a well is drilled into the center of a ground water pollution site, why might heavy pumping of this well prevent the contamination from spreading?

As you may remember from an earlier chapter, water stored in the ground is quite a large amount when compared to the amount of water in streams at any given time. This ground water is stored in rock formations known as *aquifers*. Water seeps down through the soil and thus enters an aquifer. The major types of aquifers are described in this exercise.

Sand or Sandstone Aquifer

This aquifer is made up of loose sediments, such as sand or gravel, or, if the sediments are cemented, sandstone. The aquifer begins with the water table and extends downward.

FIGURE 6-2. A sand or sandstone aquifer.

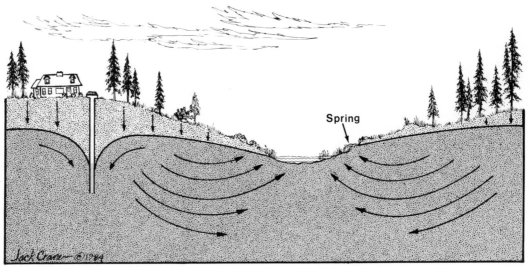

The large arrows in Figure 6–2 show the overall direction of ground water flow in this aquifer. The flow of water is in response to the force of gravity. The water travels downhill to the lowest place in the landscape. There it seeps out of the ground and feeds a stream. This is why streams flow in times without rain. They are fed by groundwater.

2. How does water enter an aquifer? Name two ways that water can leave an aquifer.

Notice on the right side of Figure 6–2 the location of a spring. A spring is found where the water table intersects the surface of the land. You probably have tasted spring water in the past. It varies in taste depending on what aquifer it originates from. This is because water dissolves very small amounts of the rock it flows through, and this dissolved material, in turn, imparts a certain distinctive taste to the water.

3. In many areas of the country, ground water must be treated before it is suitable for drinking. This involves taking out dissolved solids such as iron, sulfur, or calcium. If these dissolved solids were not in the precipitation that initially fell on the ground, where did they come from?

4. Notice in Figure 6–2 how ground water feeds into the stream. Because the stream gains water from the ground constantly as it flows downstream, it is referred to as a *gaining* stream.

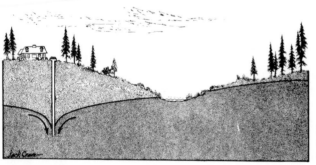

FIGURE 6–3. Sand or sandstone aquifer after heavy pumping from the well.

Figure 6–3 shows a sand or sandstone aquifer after a period of heavy pumping from the well. A stream flowing through the area could now be referred to as a losing stream. Why?

Fractured Rock Aquifer

This type of aquifer is made up of a solid rock such as granite. Unlike sand or sandstone aquifers, the rock in this type of aquifer holds little or no water. However, the rock *is* fractured and water is contained in these fractures.

36

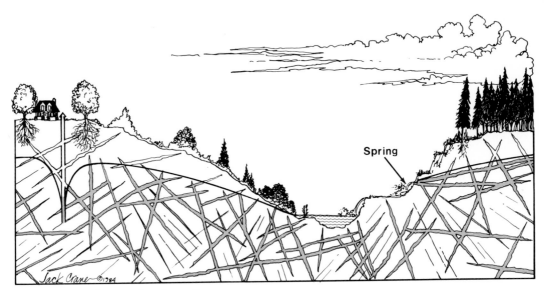

FIGURE 6-4. A fractured rock aquifer.

The overall flow of water in a fractured rock aquifer is the same as in a sand or sandstone aquifer. That is, it flows toward the streams in the area. The path of an individual droplet may be quite complicated, however, because it is restricted to flowing only in the fractures in the rock. Notice in Figure 6-4 that above the water table the fractures in the rock do not contain water while below the water table they do.

5. When drilling a well in a fractured rock aquifer, it is essential that fractures are intersected, otherwise little or no water will be obtained. In general, the more fractures that are intersected, the better the yield of the well. In Figure 6-5, one of the wells yields two gallons per minute, another ten gallons per minute, and still another fifty gallons per minute. List on your paper Well A, Well B, and Well C. Then state which well yields two, ten, and fifty gallons per minute.

FIGURE 6-5. Wells drilled in a fractured rock aquifer.

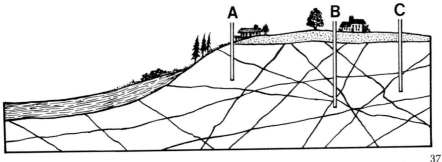

6. At a depth of several hundred feet, fractures in bedrock begin to close up. The enormous pressure of overlying rock is responsible for this. Thus, the bottom of a fractured rock aquifer may be thought of as that depth where the fractures are completely closed and cannot hold water. Where is the top of this aquifer located?

Limestone Aquifer

A limestone aquifer is similar to a fractured rock aquifer in that the water is stored in the fractures in the rock. However, the rock itself is susceptible to solution by slightly acidic ground water and, thus, the fractures have been solution-enlarged. Where large amounts of the rock have been dissolved, a sinkhole may form at the surface. Sinkholes can cause the collapse of buildings.

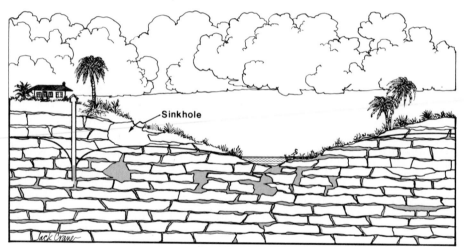

FIGURE 6-6. A limestone aquifer.

7. Why do wells generally yield more in limestone aquifers than in fractured rock aquifers?

8. Why would pollutants, such as from a faulty septic system or chemical dump, travel faster through a limestone aquifer than a sand or sandstone aquifer?

Artesian Aquifer

Ground water in an artesian aquifer is confined under pressure by a layer of rock that is impermeable to water.

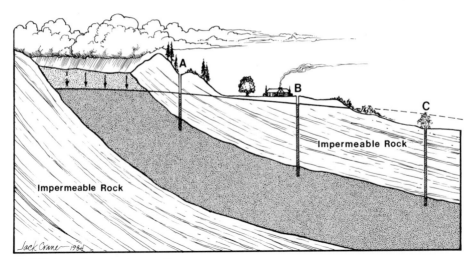

FIGURE 6-7. An artesian aquifer.

A well drilled into an artesian aquifer removes a portion of the overlying impermeable layer and allows water in the well to rise to the level that it would if the confining rock layer were not there. In some cases, an artesian well will actually flow at the surface (Figure 6–7, Well C).

Artesian water is often noted for its good quality. This is because it generally travels through a great deal of rock, which serves to have it more thoroughly purified.

9. Well drillers often keep a record of their drilling. This is called the driller's log. The driller's log for Well C in Figure 6–7 would look something like this:

Driller's log: Well C

0–4 ft	soil
4–54 ft	dense "tight" shale
54 ft	sandstone

When drilling through the shale, the well is devoid of water. What happens when the drill bit finally breaks into the sandstone?

Summary

Ground water is a natural resource that is utilized by many people, especially in areas where there is no other water supply. Enormous quantities of water are stored in the ground—much more than can be found in stream channels at any one time. This ground water is contained in geologic rock formations called aquifers. Sand, sandstone, fractured rock, and limestone can all serve as aquifers. An artesian aquifer is a special type of aquifer in which the water is under pressure.

10. Many of the techniques and strategies used in drilling for water are also used in the oil drilling business. How is oil like water?

Activities

- Read up on one of the pollutants of ground water. Become an expert on this pollutant.
- Buy a "Slushy" and suck the liquid out with a straw. How is your straw like a well in a sand or sandstone aquifer?
- Find an old hand-pump well and with the aid of a measured container such as a bucket, determine how many gallons of water you can pump out of that well in a minute.
- Build a model of an aquifer using sand. The plastic containers used for ant farms are ideal for this.
- Obtain samples of spring water from several different localities. Describe the taste of each of them in your own "field guide to good drinking water."
- Measure the flow of water out of plumbing fixtures in your home such as spigots and showerheads. What are your findings in gallons per minute?

7
THE ATMOSPHERE

The atmosphere is made up of the air we breathe. This air has a composition shown in Figure 7–1.

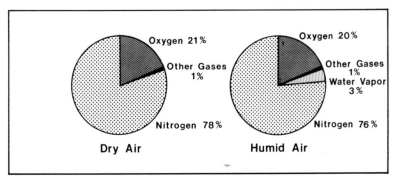

FIGURE 7–1. The composition of air. (Namowitz and Stone, 1960.)

1. List on a piece of paper the components of dry air and humid air. Then beside each gas list its percentage. What effect does water vapor have on the percentage of nitrogen and oxygen in the air?

  Most air varies in composition between dry air and humid air. The amount of water vapor in the air at any one time is referred to as its *humidity*. The more water vapor in the air, the more humid it is. The most commonly used measure of humidity is relative humidity. Relative humidity can be expressed as the amount of water vapor in the air divided by the amount of water vapor the air can hold at that temperature. Relative humidity is shown in Figure 7–2.

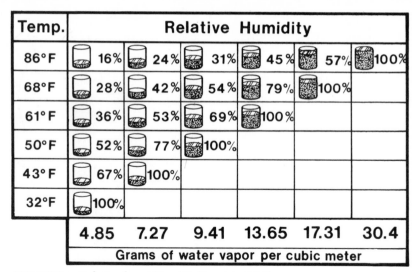

Temp.	Relative Humidity					
86°F	16%	24%	31%	45%	57%	100%
68°F	28%	42%	54%	79%	100%	
61°F	36%	53%	69%	100%		
50°F	52%	77%	100%			
43°F	67%	100%				
32°F	100%					
	4.85	7.27	9.41	13.65	17.31	30.4
	Grams of water vapor per cubic meter					

FIGURE 7–2. Relative humidity. (McNaught, 1975.)

Notice in Figure 7–2, that warm air can hold quite a bit more water vapor than cold air. For example, water at 32° F can hold only 4.85 grams of water per cubic meter, whereas water at 86° F can hold 30.4 grams/meter^3. If a cubic meter of air contained 4.85 grams of water, the relative humidity would be 100 percent if the air was 32° F. If the air was 61° F (Figure 7–2), the relative humidity would be 36 percent, and if the air was 86° F, the relative humidity would be only 16 percent.

2. A cubic meter of air is at 61° F and has a relative humidity of 36 percent. How many more grams of water can it hold?

3. A cubic meter of air at the equator has a temperature of 86° F and a relative humidity of 100 percent. This heated air will rise and in so doing will expand and cool. At an altitude of one mile above the earth's surface this air will have cooled to a temperature of about 68° F. By the time it reaches this temperature it will have given up some of its moisture in the form of rain. How much moisture will it have given up?

Our atmosphere may be thought of as a blanket of air. This blanket shields out harmful radiation from the sun and simultaneously holds in heat. The blanket is made up of several layers, the innermost of which is called the *troposphere*. The troposphere contains 75 percent of all the air molecules in the atmosphere and is responsible for the weather we have.

Other layers of air above the troposphere are much thinner. They all serve important functions, however. The ozone layer, for example, is located 12–20 miles above the earth's surface. This layer prevents harmful ultraviolet radiation from reaching the earth's surface.

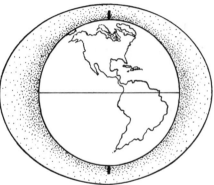

FIGURE 7-3. The troposphere, the innermost
    layer of the atmosphere. Its thickness
    is greatly exaggerated in this diagram.

Notice in Figure 7-3 that the troposphere is thicker at the equator than at the poles. This is due largely to the fact that it is much warmer at the equator and warm air expands.

Notice also in Figure 7-3 that as you go upward in the troposphere the density of air molecules decreases. This means that the air becomes thinner and cooler. It becomes cooler because there are fewer air molecules to collide with each other. It is these collisions that cause heat.

4. Snow covered peaks can be found at the equator. If the average sea level temperature at the equator is 88° F and humid air cools at a rate of 3.3° F per 1000 feet, how many feet high must a mountain be at the equator before the 32° F mark necessary for snow is reached?

As air at the equator is heated, it begins to rise. When this air reaches the top of the troposphere it divides. Half of the air travels toward the northern hemisphere and half toward the southern hemisphere (Figure 7-4).

FIGURE 7-4. Air movement at the equator.

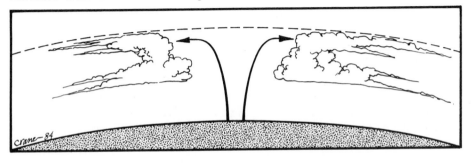

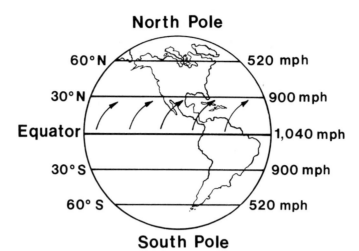

FIGURE 7-5. Speed of rotation at various latitudes.

Air traveling north from the equator gradually acquires an easterly direction. The reason for this can be seen in Figure 7-5.

If we look at the air traveling north from the equator, we see that it is traveling at the speed of 1040 miles per hour. (To an observer at the equator the air does not appear to be moving at all because the earth is also traveling at 1040 mph.) By the time this air moving north reaches 30° N latitude it has begun flowing eastward (Figure 7-5). The reason for the change in direction is that the earth at 30° N latitude is moving at a speed of 900 mph, and air traveling from the equator was initially moving at a speed of 1040 mph. Although the air from the equator has slowed down somewhat, it is still moving faster than the earth at 30° N latitude and thus "gains" on the earth by flowing toward the east.

5. On a piece of paper list the rotational speed of the earth at the equator, at 30° N, 60° N, and at the North Pole. Then figure out the distance around the world at each of these latitudes.

Air traveling north from the equator begins to cool, and at 30° N latitude it begins to sink. When this sinking air reaches the earth's surface it divides in two, with some of the air flowing north and some flowing south back toward the equator (Figure 7-6, Location A).

Air at point A, which flows back toward the equator, will soon begin to curve. The reason for this is that the air is moving with the earth at a speed of 900 mph at 30° N latitude and cannot keep up with the earth's speed of 1040 mph at the equator. Thus, to an observer standing somewhere between the equator and 30° N latitude, the wind direction will be from the northeast. These winds from just north and south of the equator to 30° N and S latitude are referred to as the trade winds (Figure 7-6). These winds were utilized by sailing vessels in the early days of trade.

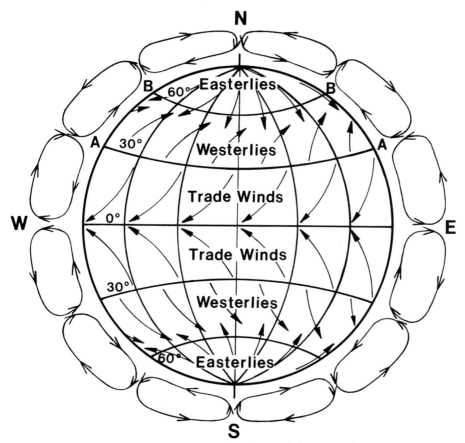

FIGURE 7-6. Surface winds and general circulation of the atmosphere.

6. Many of the famous deserts of the world occur near 30° N and 30° S latitudes. As can be seen in Figure 7-6, 30° N and S latitudes are characterized by sinking air. As this air descends to the surface of the earth it also becomes warmer. In so doing, does its relative humidity increase or decrease? Why?

7. Why is rainfall uncommon in areas of descending air at 30° N and S latitudes?

8. Notice in Figure 7-6 that half of the descending air at point A splits and travels toward the north. This air traveling north soon begins to curve toward the east. To an observer on the ground it appears as a westerly or southwesterly wind. Why does air traveling north from 30° N latitude begin to curve to the east?

At 60° N latitude air flowing south from the North Pole meets air flowing north from 30° N latitude (Figure 7-6, Location B). The two meet at the *polar front.* There warm air meets cold air NIP and the result is often storms and precipitation.

The position of the polar front varies depending on many factors, one of which is season. In summer the polar front migrates north and in winter it migrates south. The polar front is responsible for much of the weather we experience.

9. Why does precipitation often result when cold air meets warm moist air at the polar front?

Summary

The air in our atmosphere is made up principally of nitrogen, oxygen, and water vapor. The amount of water vapor present in the air is its humidity. Our atmosphere may be thought of as a blanket of air, in which the innermost layer, the troposphere, determines the weather we experience. Unequal heating of the earth causes the air in the troposphere to move. Air moving poleward from the equator tends to move toward the east because it was originally traveling faster at the equator and thus gains on the rotation of the earth. Likewise, air traveling toward the equator tends to flow to the west because it cannot keep up with the earth's spin nearer the equator.

10. The composition of the earth's atmosphere has changed throughout the planet's history. This change has usually been slow. Recent studies reveal, however, that our atmosphere has been changing more rapidly than usual since the 1800s. What do you think has accounted for this?

Activities
- Trace the flow of air in the southern hemisphere.
- Find out why condensation forms on the side of a cold drink.
- Research layers of the atmosphere other than the troposphere.
- On a map of the world sketch in where deserts of the world are located.

8 CLIMATES

This exercise will acquaint you with climates in various places of the world. By studying the climates in each of these areas, you should develop a working knowledge of climates and the factors that make them up.

FIGURE 8–1. Locations of places mentioned in this exercise.

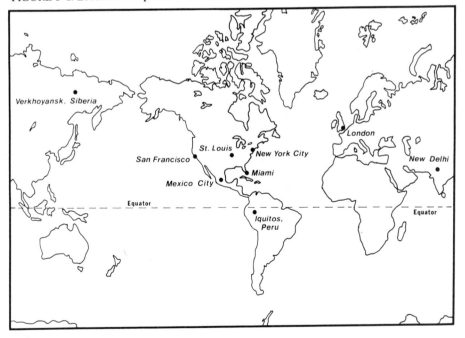

Latitude

The latitude of a place is very important in determining its climate. The intensity of the sun's rays diminishes going north and south of the tropics. This means that, in general, climates become cooler going away from the equator.

Figure 8-2 shows average temperatures and rainfalls for two cities in the world. The latitudes of these cities vary greatly. Iquitos, Peru, is located near the equator (Figure 8-1). Verkhoyansk, Siberia, is located at 68° N latitude, just inside the arctic circle. In Iquitos, Peru, the sun is almost directly overhead for much of the year; in Verkhoyansk, Siberia, the sun is never very high in the sky.

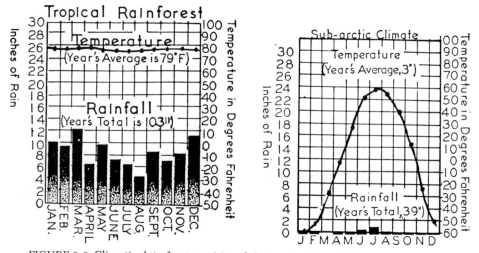

FIGURE 8-2. Climatic data for two cities: (left) Iquitos, Peru; (right) Verkhoyansk, Siberia. (Namowitz and Stone, 1960.)

1. List the months of the year. Beside each month give the average temperature for Iquitos and Verkhoyansk.

Latitude is important in determining the direction of air movement in an area. As you saw in an earlier exercise, there are areas of the world such as the equator, where the air is rising. At other areas of the world, such as at 30° N and S latitudes and near the poles, the air is sinking (Figure 7-7). As you know, rising air expands and cools, and in so doing, loses some of its ability to hold moisture. Sinking air, on the other hand, compresses and warms, and increases its ability to hold moisture. It is, therefore, less likely to give it up as precipitation.

2. Looking at Figure 8-2, give the average yearly rainfall for Iquitos, Peru, and Verkhoyansk, Siberia. Then state why you think Iquitos, a city near the equator, receives so much rainfall (*Hint:* refer to Figure 7-4).

48

Location on a Continent

Where an area is located on a continent will affect the type of climate it has. For example, St. Louis and San Francisco are two cities located at nearly the same latitude (Figure 8–1). They have the same average yearly temperature, yet their climates are quite different.

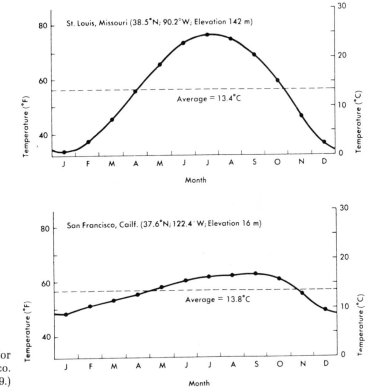

FIGURE 8–3. Climatic data for St. Louis and San Francisco. (Battan, 1979.)

3. St. Louis has a high average temperature of 75° F (in July) and a low average temperature of 34° F (in January). San Francisco has a high average temperature of 62° F (in September) and a low average temperature of 48° F (in January). What is the average temperature range for each city?

St. Louis has a higher annual temperature range than San Francisco because it is located in the middle of the continent. On a summer day when the sun is high in the sky, both cities begin to heat up. But San Francisco is located next to the ocean, and water heats up more slowly than land. Westerly winds blow across the ocean into San Francisco. These winds are cooled by the ocean, and it is these winds that keep this city from becoming oppressively hot. In winter the land cools off more quickly than the ocean. Westerly winds reflect the relatively warmer temperature of the ocean and keep San Francisco from becoming bitterly cold. Areas in the interior of the continent, such as St. Louis, do not have the buffering effect of the ocean and therefore become hotter in the summer and colder in the winter.

Prevailing Winds

Figure 8–4 shows average monthly temperatures for New York City. As can be seen in this figure, New York has hot summers and cold winters. You might expect that breezes from the nearby Atlantic Ocean would moderate the climate of this city.

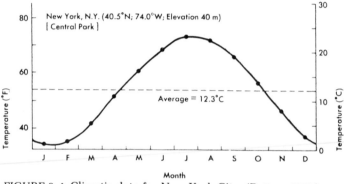

FIGURE 8–4. Climatic data for New York City. (Battan, 1979.)

4. Most of the United States is affected by prevailing westerly winds. As previously mentioned, these winds are responsible for moderating the climate of San Francisco. New York City, like San Francisco, is located next to a large ocean, but as can be seen in Figure 8–4, New York experiences hot summers and cold winters. Why does the Atlantic Ocean have little effect on buffering the climate of New York City?

Location with Respect to an Ocean Current

The climates of areas located near an ocean current are often strongly influenced by that current. The British Isles, for example, have a climate that is influenced by the Gulf Stream, a warm ocean current that flows in the Atlantic Ocean (Figure 8–5). In winter, winds blowing over the Gulf Stream bring relatively warm air to the British Isles. Thus, the Gulf Stream keeps England and Ireland warmer than what they would be normally.

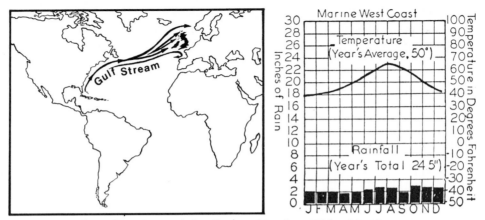

FIGURE 8-5. The Gulf Stream and the climate of London. (Right side: Namowitz and Stone, 1960.)

When warm humid air travels across a cold land surface, its ability to hold moisture decreases and a fog forms. This is what happens in the British Isles in winter, when warm winds blowing over the Gulf Stream travel across the cold land surface—fogs form.

5. London, England is particularly noted for its fogs. Why might the fogs in London be noticeably worse than in surrounding rural areas?

Location with Respect to Mountains

The seasons in India are referred to as monsoons. There are two monsoons, a winter monsoon and a summer monsoon. During the winter monsoon, winds blow south toward the Indian Ocean. During the summer monsoon, winds blow from the Indian Ocean north over the land (Figure 8-6).

FIGURE 8-6. The summer monsoon in India. (Battan, 1979.)

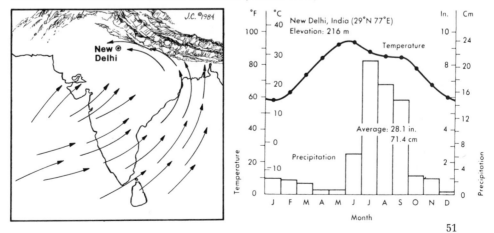

51

6. During the summer monsoon, humid air from the Indian Ocean moves north over India. When the air is forced to higher elevations by the Himalayas, it releases most of the moisture it is carrying. Why?

7. India receives more rainfall in three months than it does the entire rest of the year. In what three months does it receive this rainfall?

Altitude

Altitude is important in determining the climate of an area. Miami, Florida, for example, is located near sea level, whereas Mexico City is located more than a mile above sea level. As can be seen in Figure 8–7, these two cities have different climates.

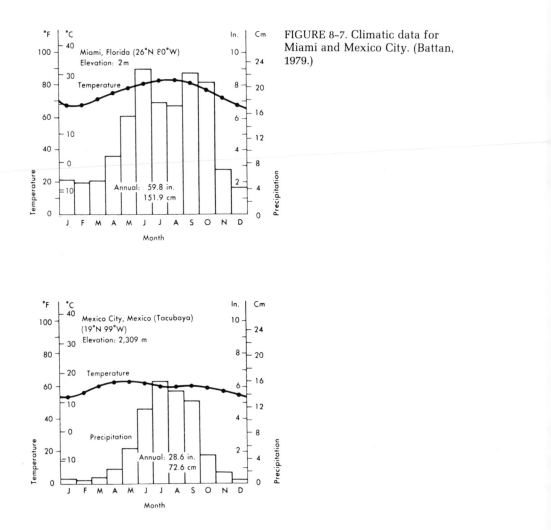

FIGURE 8–7. Climatic data for Miami and Mexico City. (Battan, 1979.)

8. Although Mexico City is closer to the equator than Miami (Figure 8-1), it has a cooler climate. Why?

Climatic Change

Climates all over the world change with time. They become warmer or colder, wetter or drier, due to slight changes in wind patterns, ocean currents, or any number of factors. Climatic change has happened in the past and will happen in the future.

9. Paintings of European landscapes that were done in the last few hundred years show people ice skating on rivers that today never freeze over. What does this say about what the climate was like in those days as compared to today?

Summary

The climate of a particular area of the world is influenced by many factors. These factors include latitude, location with respect to a land area, prevailing winds, location with respect to an ocean current, location with respect to mountains, and altitude. For any given area, any one of these factors may predominate in its influence on the climate.

Climates all over the world change with time.

10. Some areas of the world have fossils of organisms that could not possibly live in the climate in which they are discovered. Therefore, the climate must have changed since the time these organisms lived. How might continental drift cause the climate of a particular continent to become warmer, for example?

Activities
- Write a short article or essay on the climate where you live.
- Keep your own records of temperature and rainfall.
- Keep a scrapbook of weather items in the newspaper. You might want to limit your collection to just weather oddities.
- Do you know why sprinters found it easy to break world records at the 1968 Olympics in Mexico City?

9
LANDSCAPES I: WEATHERING AND EROSION

Just as a sculptor uses tools to shape a block of stone into a work of art, nature likewise shapes an area of bedrock into a landscape. It does this by first breaking down the rock at the earth's surface (*weathering*), then transporting the rock material away (*erosion*). In this exercise we will examine both weathering and erosion.

Weathering: The Breakdown of Rock at the Earth's Surface

Rock breaks down by a combination of physical and chemical processes. Physical processes cause the mechanical breakdown of rock by means such as frost wedging, abrasion, plants growing, and animals burrowing. Chemical processes cause the breakdown of rock by exposure to weak acids.

Frost Wedging. This method of weathering is based on the principle that water expands when frozen (Figure 9–1).

As seen in Figure 9–1, water must first seep into the cracks of a rock. Then, if the temperature should drop below freezing, this water will turn to ice

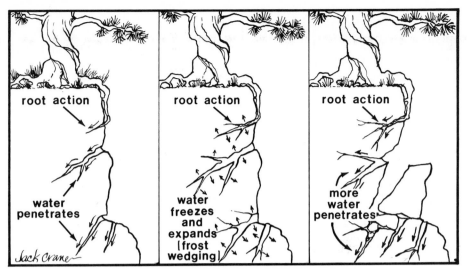

FIGURE 9-1. Physical weathering by frost wedging and root action.

and expand, thereby wedging the cracks a little further apart. Repeated periods of freezing and thawing will eventually break up the rock entirely.

Frost wedging can be quite effective in the breakdown of rock, but it is only effective in areas where the temperature fluctuates above and below freezing. It is not effective in tropical areas, for example, because the temperature seldom, if ever, drops below freezing.

1. Frost wedging is not effective in polar areas. Why not?

Abrasion. Abrasion occurs when a moving fragment of rock comes into contact with exposed rock. This happens, for instance, when the wind blows sand grains against rocks and sandblasts them. It also happens when the rock particles in a river abrade the bottom and sides of the channel. The strongest abrasion probably occurs underneath a slowly moving glacier, where ice and rock particles scour the underlying bedrock.

Plants and Animals. Plants break up rocks when their roots penetrate into the fractures and expand them. Animals such as ants, earthworms, and burrowing rodents expose new fragments of rock to the surface by their burrowing activities.

2. In the city, concrete sidewalks can be considered to be a rock. List two ways that sidewalks break up with time.

Chemical Weathering. One type of chemical weathering involves the combination of rain water with carbon dioxide in the air and soil. This water and carbon dioxide mixture forms an acid solution called *carbonic acid.* Although carbonic acid is very weak, it will, over time, dissolve and modify rock that it comes into contact with. Indeed, carbonic acid is capable of transforming large masses of granite into clay.

Another type of chemical weathering occurs from the actions of plants and animals. Plant roots and animals secrete acids that attack rocks chemically.

3. Why is chemical weathering generally more effective in a humid climate than in a dry climate such as a desert?

4. The physical and chemical weathering processes described here usually work together to break down rock at the earth's surface. Chemical weathering, for example, can only operate on the outer surface of a rock. But with the help of frost action, chemical weathering becomes much more effective. Why?

Erosion: The Transportation of Weathered Rock

Once rock has broken down at the earth's surface, it can then be moved. Agents that move this material are gravity, streams, glaciers, and wind. Let's examine each of these agents of erosion.

Gravity. Avalanches and landslides are spectacular examples of how gravity can move rock material. The majority of weathered rock, however, is moved by much slower processes, such as *creep* (Figure 9–2).

FIGURE 9-2. Creep on a hillside. (Fagan, 1965.)

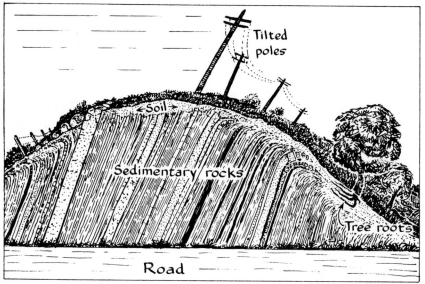

Notice in Figure 9–2 that broken rock and soil near the top of the ground has "crept" downslope somewhat. This is due to the long slow pull of gravity.

5. How might the tombstones in an old cemetery reflect the process of creep?

Another way that gravity pulls rock material downslope is by *slump*. Slump occurs where a slope of material has become too steep to support itself (Figure 9–3).

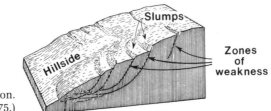

FIGURE 9–3. Slump formation. (Fagan, 1975.)

6. Why might heavy rains accelerate the formation of slumps?

Streams. Streams are one of the most important agents for moving rock material. They carry weathered rock that has been brought to them by gravity processes such as slump and creep. In addition, in upland areas they actively erode the landscape (Figure 9–4).

FIGURE 9–4. Stream erosion in an upland area. (Fagan, 1965.)

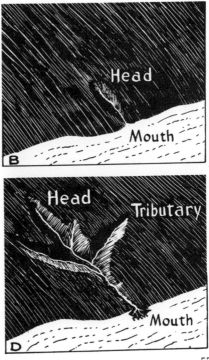

7. Why might the downward cutting of a stream channel cause slumps to form?

Glaciers. Glaciers are sheets of ice which slowly move due to recrystallization within the ice. This movement is usually on the order of only a few centimeters a day.

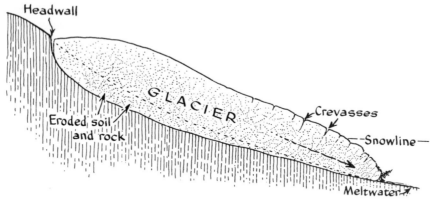

FIGURE 9–5. The movement of a glacier. (Fagan, 1965.)

Glaciers not only move weathered rock, they also *create* weathered rock by scouring the bedrock over which they move.

8. Glaciers are capable of transporting giant boulders, whereas streams are not. Why is this?

Wind. The wind generally plays only a minor role in erosion. It has its greatest effect in desert areas, where it can pick up loose rock material and transport it.

In the 1930s, the wind was responsible for eroding topsoil from farms in much of the midwest United States. Poor farming practices together with severe drought caused this area to be known as the "Dust Bowl."

9. As an area becomes drought-stricken, why does the erosive power of the wind increase?

Summary

Weathering and erosion are the "tools" that nature uses to shape a landscape. Weathering is the breakdown of rock at the earth's surface by processes such as frost wedging, abrasion, plants growing, animals burrowing, and chemical action. Erosion is the removal of weathered material by means of gravity, streams, glaciers, or wind.

10. Soil forms wherever weathering breaks down rock faster than it can be removed by erosion. Would you say this is a common occurrence?

Activities

- Take a field trip and observe weathering and erosion processes firsthand.
- Put some hydrochloric acid on a piece of limestone, then describe the chemical weathering that takes place.
- Why does frost action break a bottle of soda that is left in the freezer?
- In a series of drawings, show ways that weathering processes interact.

10

LANDSCAPES II:
THE INFLUENCE OF CLIMATE

The type of landscape that forms in an area is due largely to the climate in that area. Climate determines how fast weathering and erosional processes operate. It also determines *which* weathering and erosional processes will operate. In this exercise we will look at how landscapes form under different climates.

Tropical Landscapes

Tropical landscapes are located near the equator. Many of these landscapes can be considered to be rainforest—they receive rainfall on the order of eighty inches per year. This heavy rainfall supports a lush growth of vegetation. (Figure 10–1).

FIGURE 10–1. Tropical rainforest areas of the world. (*World Book*, 1984.)

1. The weathering of rock by chemical action is more intense in a tropical rainforest than in other landscapes. Why?

2. The Amazon River in South America drains an area of tropical rainforest. This river carries the largest volume of water of any river in the world. Part of the reason for this is that it drains a large area of land; but even so, the Amazon still carries an unusually large amount of water. What is the other reason that this river carries so much water?

Temperate Landscapes

Temperate landscapes form in areas where the climate is neither exceedingly hot nor exceedingly cold. These landscapes are generally located between the tropics and the polar circles. They experience a distinct summer and winter and they have a moderate amount of rainfall.

Rock weathers by several methods in a temperate landscape. It weathers by frost action, by plants growing, animals burrowing, and by chemical action.

3. The intensity of weathering processes varies according to season in a temperate landscape. How would frost action vary from one season to another in a temperate landscape?

4. Chemical action speeds up markedly in warm temperatures, as does the activity of plants and animals. The increased growth of plants, for example, can be seen in tree rings. If you have ever looked closely at tree rings, you will notice that there is such a thing as summer growth and winter growth. Looking at one individual ring, the summer growth part is typically much wider than the (darker) winter growth part. This is because the plant does most of its growing in the summer months. If this growth pattern of the tree trunk holds true for the tree roots also, why would tree roots be more effective in breaking apart rock in the summer than in the winter?

Erosion in temperate landscapes takes place by streams and by gravity processes such as creep and slump (Figures 9–2 and 9–3). Streams, in particular, play a major role in shaping these landscapes. They carry away weathered material and transport it elsewhere, usually far away.

5. Much of the material carried by a stream is deposited when that stream enters a relatively calm body of water such as a lake, a bay, or the ocean. What causes the stream to deposit most of the material it was carrying at such locations?

Polar Landscapes

Polar landscapes form in areas where the climate is cold for most, if not all, of the year. These landscapes are found near the north and south poles and in high mountainous areas of the world. They are characterized by little or no vegetation and, in some areas, ground that remains frozen all year long (permafrost).

Many polar landscapes are covered by glaciers. Glaciers form where snowfalls do not melt, even in summer. As successive snowfalls build up, the bottom snow layers are compressed into ice. This ice, when it gets to be several hundred feet thick, slowly begins to flow. It flows due to recrystallization of ice which is still in its solid state. This flow is too slow to be seen, but nevertheless it does occur.

Much abrasion takes place at the bottom of a glacier. The ice literally gouges out rock and leaves what is known as a *U-shaped valley*.

6. Figure 10–3 shows two valleys. Both valleys now have a stream flowing in them. However, one of them was at one time occupied by a glacier. Which one was it and why?

FIGURE 10–2. Glaciers.

FIGURE 10-3. Two valleys.

As a glacier moves downhill, it carries with it a lot of loose rock. This rock ranges in size from as large as boulders to as small as dust. Once this material reaches the end of the glacier it is deposited on the ground. Water melting from the glacier then picks up much of this rock mixture and moves it downhill.

7. Large rocks and boulders are often found close to the bottom end of a glacier while smaller rock material, such as sand and gravel, is found farther downslope. Why would glacial meltwaters tend to carry the small material downslope but not the large material?

Arid Landscapes

Arid landscapes occur in areas where there is little rainfall. These are desert and semi-desert areas, and they occupy about one-third of the world's land.

FIGURE 10-4. Arid areas of the world. (Meigs, 1956.)

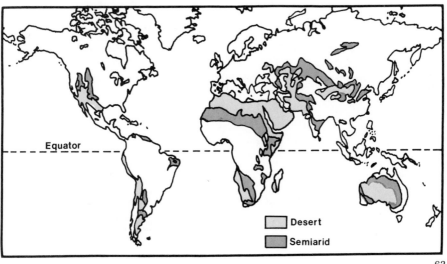

The rain that does fall in arid areas often comes during heavy *cloudbursts*. When a cloudburst hits, the water runs off the land quite rapidly and carries with it weathered rock material.

8. If a cloudburst were to strike both an arid landscape and a temperate landscape, why would the erosion be greater in the arid landscape?

Running water is the main sculpturing tool in arid areas. Between rainfalls, the wind also sculptures the landscape, although its overall effect is not as great as that of running water. The wind blows sand grains around and in some areas piles them into dunes. Particles that are smaller than sand may be picked up by the wind and removed entirely from the desert environment.

Coastal Landscapes

Coastal landscapes occur where the ocean meets the land. Running water in the form of waves is the main sculpturing agent of these landscapes. In some areas waves break down rock and carry it out to sea; in other areas beaches are formed. The type of coastline that will form in any one area is determined largely by the strength of the rock in that area and the intensity of the waves that strike the area.

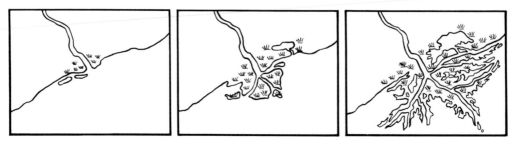

FIGURE 10-5. Formation of a delta. (Namowitz and Stone, 1960.)

A delta is a type of coastline that builds out into the ocean. It forms at the place where a river enters the ocean. The rock material that the river was carrying is dropped when the river's current is lost in the relatively calm ocean waters.

9. Why don't deltas form where wave action is intense?

Summary

The type of landscape that forms in an area is determined largely by the climate of that area. This is because climate controls the weathering and erosional processes, which are the tools that shape a landscape. Tropical

64

landscapes, temperate landscapes, glacial landscapes, and arid landscapes all have developed the way they have largely because of the climate. Coastal landscapes are also affected by climate, although factors such as rock type and wave action are often more important in determining the appearance of these landscapes.

10. List the five types of landscapes mentioned in this exercise. Beside each landscape list its main agent of erosion (gravity, running water, ice, or wind). One of the landscapes has two major agents of erosion.

Activities

* Sketch a landscape in a local park.
* Make a papier-mâché model of a polar landscape.
* Describe the changes in the countryside closest to your home in each season of the year. How does nature wear down the landscape each season? An interesting way to record these changes is to take a series of photographs in the same spot for thirteen consecutive months.
* Describe in a series of drawings or cartoons how a landscape wears down over a period of years.

SOILS

As an area of rock becomes exposed at the earth's surface, it begins to disintegrate. Physical and chemical weathering processes attack the rock and split it into smaller fragments. The smaller fragments are then broken down into yet smaller fragments, until eventually a fine-grained layer of sediment is formed. Microorganisms such as bacteria and fungi soon find their way into this sediment layer and begin their life processes. Then animals such as earthworms and ants find the habitat suitable, as do small plants. As these organisms live and die they give an organic component to the layer of sediment, and soil is formed.

Soil Horizons

Soils can be divided into three horizons or layers. These layers are referred to as the A horizon, B horizon, and C horizon (Figure 11–1).

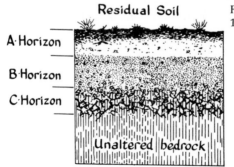

FIGURE 11–1. Soil horizons. (Fagan, 1965.)

66

A Horizon. This layer is the *topsoil* or root zone. It usually contains *humus* (partially decayed plant and animal tissue) and is usually the most fertile of the three horizons.

B Horizon. This layer is considered the *subsoil.* It contains weathered mineral matter but no humus. It also contains many nutrients and clay particles that were leached or washed down from the A horizon.

C Horizon. This layer lies directly over bedrock and grades into *bedrock.* It consists of pieces of detached bedrock mixed with weathered material.

1. Soil horizons that develop in steep areas are generally not nearly as thick as those on flat land or gently sloping areas. Why would water running off a steep slope erode the soil faster than an equal amount of water running off a gentle slope?

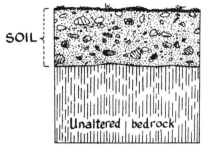

FIGURE 11-2. A transported soil. (Fagan, 1965.)

2. Figure 11-2 shows soil that did not develop in place, but rather, was transported there. A landslide, stream, or glacier may have transported the soil to its present location. Would you expect rock fragments in the C horizon of such a soil to match the rock type of the underlying bedrock? Why or why not?

3. Animals such as earthworms, ants, and rodents rework soil by burrowing through it. This reworking aerates and mixes nutrients in the soil and thus is very valuable. Would these organisms tend to make the boundaries between the A, B, and C horizons more distinct or less distinct? Why?

4. Many plants draw nutrients through their roots from the B horizon of the soil. The nutrients they draw were initially leached downward out of the A horizon. These nutrients are used in forming the structure of the plant. Thus, many plants may be thought of as nutrient "pumps" because they pump nutrients from the B horizon and eventually return them to the A horizon. If these nutrients are stored in the structure of plants, how are they eventually returned to the A horizon?

Microorganisms

Microorganisms such as *bacteria* and *fungi* play a very important role in soil. They prepare soil nutrients in such a form that they can be taken up and used by plants. Other microorganisms decompose dead plants and make them part of the soil once again.

In the process of decomposition, humus is formed. Humus commonly forms a rich organic layer on top of the soil. Dead plant and animal material is constantly being added to the top of this humus layer. At the bottom of the layer, decomposition is constantly absorbing the humus into the soil.

5. Microorganisms decrease their activity in colder temperatures and increase it markedly in warmer temperatures. This is the reason that in some arctic areas peat bogs are formed. In such areas microorganisms cannot keep pace with the dead plant matter being deposited on the ground and, thus, layer upon layer of humus builds up to form a deposit of peat.

    In tropical areas the vegetation is often fast growing and very lush. In such areas you might expect to find a lot of humus in the soil. Yet most tropical soils have very little humus in them. Why?

6. Soil is composed of minerals, humus, living organisms, air, and water. When an area is permanently flooded, such as when a dam is built, trees standing under water will soon die. What is lacking in the soil that causes this?

Soil Conservation

Much soil that is under cultivation is being eroded at an alarming rate. Figure 11-3 shows a common way that this soil erodes.

Notice in Figure 11-3 that the impact of a raindrop on unvegetated land serves to loosen the soil. Because the soil is loosened, it is more likely to wash away as the rain continues. With time, gullies form on the landscape.

At the beginning of a crop year, soil is plowed and gullies disappear. But soil erosion has nevertheless occurred. There is not as much topsoil present as there was the year before. As the years go by, the topsoil is gradually eroded away and the plow brings subsoil (B horizon) to the surface. This subsoil is generally not as fertile as the original topsoil, and crop yields may thereby decline. The whole process is gradual, and farmers may not recognize it at first. The use of improved fertilizers also tends to mask the effect of eroding farmland. With fertilizers, farmers may continue to get high yields from their crops and may think that their land is still in very good shape when in fact it may be eroding to dangerous levels.

FIGURE 11-3. The erosion of topsoil. (USDA—Soil Conservation Service.)

7. Soil formation is usually a very slow process. The weathering of bedrock may form an inch of soil in 1500 years. Soil erosion, on the other hand, often removes topsoil at a much quicker rate. For example, in areas where soil conservation is not practiced, erosion may remove as much as one inch of topsoil in fifteen years. In such an area, soil is being eroded how many times faster than it is being formed? Also, if such an area has six inches of topsoil, how many years will it be before this layer is eroded away?

Steps can be taken to minimize soil erosion. One method is contour plowing, in which furrows are made around a hill instead of up and down the hill. This practice helps retard water from flowing rapidly down the hill and eroding the soil. Another practice is *mulch tillage,* in which most of the plant of the crop is left after the crop has been harvested (Figure 11–4).

Notice in Figure 11–4 that corn plant mulch has been left on the ground only in the field on the right. This plant mulch will protect the ground in several ways. One is that it will break the fall of raindrops which would otherwise loosen the soil. Another way is that it will retard the flow of water off the field. Individual leaves and stalks will act as small dams which temporarily restrict the flow of water. By so doing, they slow down the speed of the runoff water and thereby decrease its ability to erode soil. Plant mulch also soaks up moisture which will be released to the soil when the mulch is plowed under the following spring.

FIGURE 11–4. Conventional farming versus mulch tillage. (USDA—Soil Conservation Service.)

8. In Figure 11–4, what evidence is there of erosion on the field on the left that is not evident on the field with mulch on the right?

Many farmers use plant mulch for animal feed and are reluctant to leave it on the field. Thus, the farmer on the left field in Figure 11–4 will save money on animal feed this year while the farmer on the right will not. But the farmer on the right will need less fertilizer the following year and will not have nearly as much of his soil eroded. In the long run he will be in better financial shape.

9. What effect does soil erosion have on stream pollution?

Summary

The physical and chemical breakdown of rock at the earth's surface initiates the formation of soil. Soil forms into three horizons—an A horizon, a B horizon, and a C horizon. Burrowing animals rework soil, and microorganisms such as bacteria and fungi play important roles in the growth and decay of plants in the soil.

Soil conservation is a practice that is used to keep topsoil from eroding off cultivated land. One method of soil conservation is mulch tillage. Here the remains of crops are left on the field after the crop itself has been harvested. These plant remains help protect the soil from erosion.

10. A new method of farming being used in some areas today is called *no-till farming*. With this method the soil is not actually turned over as it is in conventional farming. Instead, thin troughs are dug in the soil and the seeds are planted in these troughs. Natural vegetation is allowed to remain between troughs although it is controlled with herbicides.

No-till farming conserves the soil better than even mulch farming. With mulch farming, erosion often occurs after the crop has been planted in the spring but before it has grown very large. The mulch is not very helpful at this time because its location with respect to the soil has changed. Where is the plant mulch after the spring plowing?

Activities

- Observe the soil profile in a road cut or an excavation. Describe each soil horizon and measure its depth. Take a picture of it.
- Observe several farm fields and note what soil conservation measures are being used or what measures could (or should) be used.
- Visit or call your local Soil Conservation Service. They have many useful pamphlets.
- Find out what "prime" farmland is and why it should be conserved.

12 THE HISTORY OF THE EARTH

About 4.6 billion years ago, a cloud of gas and dust particles condensed to form the earth (Figure 1–2). Early in its history, the earth contained a hodgepodge of materials, among which were radioactive elements such as uranium, thorium, and potassium. With time, these radioactive elements gave up heat that was sufficient to melt large portions of the planet. When this happened, pockets of iron that were scattered throughout the earth gradually sank to the center and formed the inner and outer cores.

1. What property does iron have that would allow it to sink through other material?

About 3.8 billion years ago the earth had cooled sufficiently that some pieces of solid crust that formed were not destroyed at a later time. Molten material and gases continued to erupt from beneath the surface, however. The gases contained large amounts of water vapor, which condensed in the atmosphere and fell as rain. The earth was still so hot that this rainfall instantly

FIGURE 12–1. Early history of the earth. (Fagan, 1965.)

72

turned to steam upon striking the ground. With the passage of time, however, the crust of the earth had cooled enough so that water could gather on the surface. Eventually, enough water accumulated to form the vast oceans.

2. Small amounts of water are being added to the hydrologic cycle each year. How?

Origin of Life

At about the same time the oceans were forming, disturbances in the atmosphere were generating spectacular lightning storms. Lightning, together with ultraviolet radiation from the sun, was responsible for breaking apart molecules in the atmosphere and combining them into more complex molecules. These newly formed molecules found their way into the oceans and gradually became complex enough so that one day a molecule was formed that was capable of reproducing itself. It is believed that this was the beginning of life on the earth.

Organic molecules are the building blocks of life. These molecules can be generated in a glass vessel by mixing together the gases methane, ammonia, hydrogen sulfide, and water vapor. Electric sparks are then shot through the vessel, and after a few hours a brown tar is formed that contains many organic molecules (Figure 12–2).

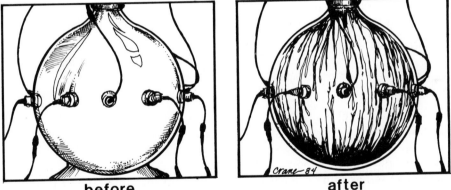

before   **after**

FIGURE 12–2. Manufacture of organic molecules in the laboratory: (left) before; (right) after. (Courtesy of Bishun N. Khare and Carl Sagan, Cornell University.)

3. The purpose of the experiment shown in Figure 12–2 was to re-create the conditions that existed on the primitive earth to see if the building blocks of life could have been generated in such an environment. The gases methane, ammonia, hydrogen sulfide, and water vapor corresponded to the gases that probably composed the early atmosphere. What occurrence did the electric sparks that were shot through the gas mixture correspond to in the early atmosphere?

Early Life

Life on earth is thought to have begun about 4 billion years ago. However, only one-celled organisms such as algae inhabited the earth for the following 3 billion years.

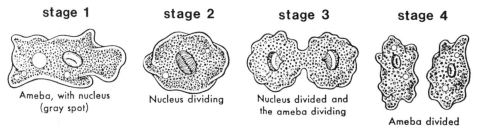

FIGURE 12–3. Reproduction in a single-celled organism. (*World Book*, 1984.)

It was not until about 1 billion years ago that multicelled organisms first began to evolve. One theory on how such life began is that a mutation in a particular cell may have prevented that organism from dividing after it had reproduced. Thus, the organism would have remained in stage 3 of Figure 12–3.

As the two cells stayed together, the one may have been better at moving about, and the other may have been better at ingesting food. Thus, each cell had its function, and the two cells so joined were more efficient than the single-celled organisms around them. The increased efficiency made that organism's chances of survival greater and thus increased the likelihood that this mutation would be passed on to future generations.

As organisms with many cells began to evolve, each of the cells in an organism became further specialized. Cells began to work together as a team, with each cell or group of cells (organs) having its own particular function.

The Time Scale

About 600 million years ago, evolution began in earnest, with a multitude of life forms appearing rapidly in the fossil record. Figure 12–4 shows the history of the earth divided into four distinct eras. As seen to the right of this figure, these divisions are based primarily on the life forms that inhabited the earth during each of these times.

Notice in Figure 12–4 that the Precambrian was by far the longest era but it also had by far the simplest life forms.

4. How many years did each of the four eras last?

5. In what era did the first fish appear?

6. The first land plants appeared in what era?

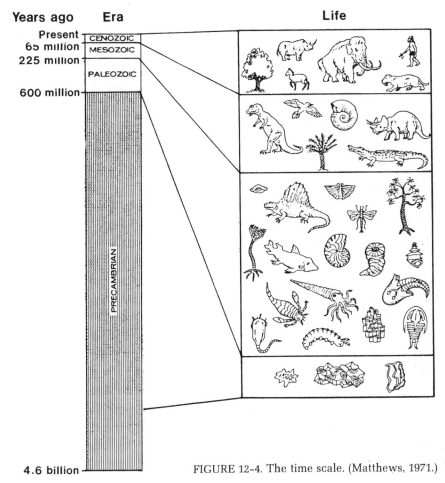

Years ago	Era	Life

Present
65 million
225 million

CENOZOIC
MESOZOIC
PALEOZOIC

600 million

PRECAMBRIAN

4.6 billion

FIGURE 12-4. The time scale. (Matthews, 1971.)

7. The origin of life occurred in what era?

8. During what era did the dinosaurs inhabit the earth?

9. The human species appeared on the earth only a few million years ago. We as a species made our appearance late in the Cenozoic era. Thus, we may be thought of as newcomers to the planet Earth.

It is possible that the human species has been around for 10 million years. If the age of the earth is 4.6 billion years, what fraction of that time has mankind been on the earth?

Paleozoic Life

In the early Paleozoic Era, life had not yet evolved on land. It was abundant in the seas, however, as can be seen in Figure 12-5.

FIGURE 12–5. Life in an early Paleozoic sea. (Smithsonian Institution.)

10. List the five types of organisms that are labeled in Figure 12–5, then state what organism the giant Cephalopod in the foreground is about to feed on.

By mid-Paleozoic time, plants had begun to colonize the land. Trees that lived during this time resembled giant ferns (Figure 12–6). Insects and arachnids such as spiders also evolved during the middle of this era.

FIGURE 12–6. A landscape in mid-Paleozoic time. (Courtesy of the New York State Museum.)

11. Figure 12-6 is a reconstruction of a mid-Paleozoic landscape. Notice that this reconstruction does not show any large animals. Similar reconstructions of this time period may show insects or spiders, yet they also do not show large animals of any kind on the land. Why not?

In late Paleozoic time, large land areas were covered by swamps. The trees that grew in these swamps looked quite different from those of today. A wide variety of animals had taken to the land by this time, and some insects, such as one that resembled today's dragonfly, had a wingspan of over two feet!

FIGURE 12-7. A landscape in late Paleozoic time. (Peabody Museum, Yale University.)

12. Generations of plants that lived in the swamps of late Paleozoic time built up a thick layer of humus. With time and the pressure of overlying material, this layer of humus compacted into peat, then lignite, and finally coal. Why is coal referred to as a fossil fuel?

Mesozoic Life

Many organisms became extinct at the end of the Paleozoic Era, 225 million years ago. It is believed that intense mountain building along with volcanism changed climates dramatically and many organisms could not adapt. Those that did adapt survived into the Mesozoic Era and evolved into life forms such as those shown in Figure 12-8.

13. What is the name of the large creature in the center foreground of Figure 12-8?

FIGURE 12–8. A landscape in the Mesozoic Era. (Peabody Museum, Yale University.)

14. Triceratops (left foreground) had what sort of protection for defending itself against attackers?

15. What evidence is there that the atmosphere in this area is changing?

Cenozoic Life

About 65 million years ago, mass extinctions again occurred on earth. Perhaps the best known casualties of these extinctions are the dinosaurs. All of the dinosaurs became extinct at this time. Recent evidence suggests that the extinctions were caused by extreme climate changes that resulted when an asteroid collided with the earth. Whatever the cause, many plant and animal species were unable to adapt and died out. Those species that survived lived to inherit the earth during the Cenozoic Era.

16. Figure 12–9 shows a landscape in early Cenozoic time in the Rocky Mountain states. Notice that many of the animals resemble modern-day animals to some extent. Some of these animal species became extinct and some evolved into the species we see on earth today. What evidence do you see in Figure 12–9 that the climate was warmer in this area than it is today?

FIGURE 12-9. A landscape in Cenozoic time. (Peabody Museum, Yale University.)

Continental Drift

Continental drift has probably occurred during much of the history of the earth. Figure 12–10 shows the locations of the continents at different periods of time.

17. Many organisms that freely roamed from North America to Africa in late Paleozoic time are not able to do so today. What happened between that time and today to cause this?

18. Why was the climate of North America probably warmer in late Paleozoic time than it is today?

FIGURE 12-10. Continental drift.

LATE PALEOZOIC
225 m.y. ago

LATE MESOZOIC
65 m.y. ago

LATE CENOZOIC
Present Day

Inland Seas

Shallow continental seas have covered large land areas in the past (Figure 12–11).

Evidence for inland seas is found in rock layers that contain fossils. These fossils are of organisms that could have only lived in the sea.

19. Name three states that were completely covered by water in late Mesozoic time.

FIGURE 12–11. Inland seas in late Mesozoic time. (top: after Z. Burian, Praha; bottom: Dunbar, 1960.)

Summary

The earth is 4.6 billion years old. Its early history was characterized by extensive volcanic eruptions and a dense steamy atmosphere. With time, the planet cooled sufficiently for oceans to form. Life began in these oceans about 4 billion years ago. However, only very simple organisms inhabited the earth until about 600 million years ago, when the Paleozoic Era began. Life evolved rapidly in the Paleozoic Era as organisms occupied the seas and evolved to occupy the land as well. In the Mesozoic Era, some of the organisms, especially dinosaurs, reached mammoth proportions. Mass extinctions ended this era and heralded in the Cenozoic. Cenozoic life saw the evolution of the plants and animals we see today.

The history of life on the earth must be viewed against a backdrop of drifting continents; it must also be kept in mind that inland seas have covered large land areas in the past.

20. The Paleozoic Era ended when mountain building and volcanism changed climates and resulted in mass extinctions. The end of the Mesozoic Era also saw extinctions on a large scale due to rapid climatic change. Mankind has the power to bring the Cenozoic Era to a close. How?

Activities

- Research an organism such as the trilobite and trace when it lived and when it became extinct.
- Read up on the dinosaurs. Many books are available and new ones are published each year.
- Draw a landscape or seascape of what life looked like at a particular time in the earth's history.
- Discuss ways a particular species could become extinct.

13

EVOLUTION

Figure 13–1 shows some of the many kinds of domesticated dogs that have been produced by breeding.

All of the dogs shown in Figure 13–1 were descended from a single type of wild dog, which was probably the wolf. As individual dogs were domesticated, selective breeding was done with them. That is, instead of allowing individuals to breed with the mate of their choice, as is the case in the wild, only those dogs with certain characteristics were allowed to breed. If large dogs were desired, for example, a large male and female were chosen to breed.

FIGURE 13–1. Domesticated dogs. (Moore, 1958.)

It was thought (and correctly so) that the offspring of such a match-up would most likely grow to be larger than average. Generations of such breeding succeeded in producing large dogs such as today's Great Dane or mastiff.

1. Some dog breeders desired features in a dog other than large size. Some breeders may have desired a small dog with a long snout, for example. Such dogs are useful for flushing small animals out of their underground burrows. As these breeders bred the newly domesticated wild dog, what characteristics would they look for in a dog that was to be bred? Name a modern-day dog that is the result of such breeding.

When we look at the wide variety of dogs there are today, it may seem hard to realize that all these breeds can trace their descent to a common ancestral wild dog—yet this is the case. Artificial selection of certain individuals for breeding in each generation has led to the wide differences we see in dogs today.

2. Artificial selection has resulted in many of the crops we currently grow. Many crops, such as corn, have been bred to such an extent that they can no longer survive in the wild by themselves. The mechanism for scattering corn seeds, for example, has been lost in the process of breeding for larger ears of corn. Large ears of corn mean a higher yield to the farmer. Whether the corn plant can scatter its seeds from year to year in the wild is not a concern to farmers. Why not?

The Theory of Evolution

In 1859 Charles Darwin, an English biologist, proposed the theory that *all living organisms on earth were descended from a common ancestor.* Just as today's breeds of dogs can be shown to have originated from a common wild dog, Darwin proposed that all life on earth originated from a common ancestor. This ancestor was a one-celled microscopic organism. How Darwin's theory relates to the animal kingdom is shown in Figure 13–2.

Figure 13–2 shows many organisms that are alive today as well as some extinct species such as trilobites. According to Darwin's theory of evolution, an animal's ancestors can be determined by tracing back on the tree of life. By doing this we can see that the early ancestors of reptiles, for example, were amphibians.

3. According to Figure 13–2, what were the early ancestors of the amphibians?

4. What is the name given to the group of one-celled organisms from which Darwin believed all life orginated?

Darwin formulated his theory of evolution based on three observations. Let's examine each of these.

FIGURE 13-2. The tree of animal life. (© Paleontologisk Museum, University of Oslo.)

Differences Between Individuals. Darwin recognized that all individuals of a species differ from each other. In people we can see differences in body size and build, color of eyes, hair color, etc. In plants there are differences in such things as height, the overall shape of the plant, and the number of seeds produced. For any given plant or animal species, there are differences that exist between individuals in that species.

Tendency Toward Overpopulation. Every species on earth produces enough offspring such that, if a high percentage of these offspring lived, that species would soon overrun the earth. Many trees, for example, produce thousands of seeds each year. For any given type of tree, if each of its seeds were to grow into mature trees, which in turn produced seeds, it would not be long before that species of tree was predominant on the earth. Another example is fish, which commonly lay millions of eggs at a time. If each of these eggs were to grow into a mature fish, that species would soon dominate the seas. Even organisms that produce only one offspring per year. such as horses, are capable of overpopulating the earth. One female is capable of producing a colt each year for many years.

Struggle for Survival. Darwin recognized that there are limited resources on the earth. There is a limited amount of food, for example, just as there is a lim-

84

ited amount of space on land and in the sea. Because of limited resources, all organisms must compete for the resources that are available. This creates a struggle for survival in which those that are most fit survive.

Natural Selection

In looking at the struggle for survival between organisms, Darwin stated his belief that it was the environment that determined which individuals would reach maturity and produce the next generation. This principle he called *natural selection.* To illustrate the principle of natural selection, let's look at amphibians. Amphibians are those animals, such as frogs, that spend part of their lives in water and part on land.

According to the theory of evolution, the ancestors of amphibians were fish. At times in the past history of the earth, shallow inland seas covered much of what today is land. As these seas began to dry up, they formed lakes. These lakes in turn began to dry up, which meant that fish that inhabited them were in danger of extinction. Among those fish were certain individuals that had both gills for breathing in water and primitive lungs for breathing out of water. Some of these fish had well-developed front fins as well and were able to crawl out of water onto the land. Once on land they encountered a whole new food supply of insects and plants. Thus, those individual fish that could make it up on land were assured of a food supply and were more likely to reproduce than other members of the species that were not able to crawl up on land. The offspring of these stronger fish were more likely to have well-developed forelimbs also, and with time, these organisms ceased to be fish any longer and instead resembled amphibians (Figure 13-3).

FIGURE 13-3. Evolution of the first amphibians. (Courtesy of the Library Services Department, American Museum of Natural History.)

5. Natural selection is not always a tooth-and-claw fight between individuals. The term *survival of the fittest* is sometimes used, but who is most fit varies greatly and sometimes is determined by those individuals who can get along the best. Most apes, for example, live together in groups. This increases an individual's chances for survival in many ways. Name a way you think this would increase its chances for survival.

Mutation

Darwin thought of evolution as a very slow process, in which new species gradually arose due to natural selection in a changing environment. Later researchers found that there is a process by which species can change more rapidly. This process is known as *mutation*. In any given species, there are occasionally individuals born with mutations. Frogs, for example, have been born that have had four back legs instead of two. This mutation, as is true with most mutations, greatly decreased the individual's chances for survival and, thus, did not find its way into later generations. Occasionally, however, a mutation occurs that increases an organism's chances for survival. This may have happened with an amphibian that was born with the ability to produce eggs that could be laid on land. Such an individual would not have had to return to water to lay her eggs. If this mutation was passed on to succeeding generations, the result was the evolution of an organism that no longer was dependent on water for part of its life cycle. Such an organism could travel further inland and occupy new habitats. This newly-evolved organism would have been a reptile.

Mutations provide the raw material for evolution. Differences between individuals of any given species can be traced back to major or minor mutations in their ancestors. It is natural selection that then determines which mutations will survive and be passed on to future generations.

6. Severe mutations occasionally occur in the human population. Individuals born with genetic defects, as they are called, may or may not be able to lead normal lives. Why might such individuals choose to talk to a genetic counselor before having children of their own?

Similar Bone Structures

If evolution has indeed occurred in the past, we should expect to see similarities in structure between species that are close to each other on the evolutionary tree. Figure 13–4 shows a series of skeletons from reptiles to man.

Notice in Figure 13–4 that many features are similar between the chimpanzee and man. Arm and leg structures, for example, are remarkably similar, as are rib cage and backbone structures.

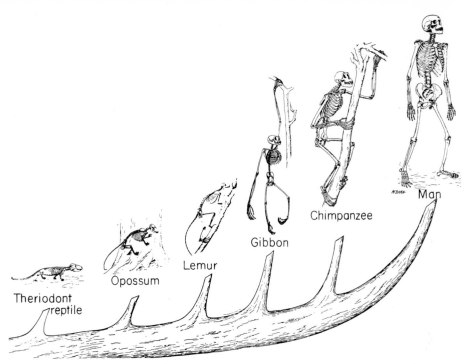

FIGURE 13-4. Skeletons. (Courtesy of the Library Services Department, American Museum of Natural History.)

Scientists believe that man's early ancestors were apes, such as the chimpanzee. These apes lived in forest areas, for the most part. With time, some of them began to move out into grassland areas. Among those apes that did move into the grasslands were those individuals that had a more upright posture than others. These individuals found that they could move around better. This greatly increased their chances for survival. Because these individuals tended to survive, they were able to breed and pass on their upright posture to future generations. With time the apes that lived in the grasslands became fully upright and acquired other characteristics that suited them better for life in their grassland habitat. With time they no longer looked like apes; instead they looked like early man.

7. In Figure 13-4, what is the major change that can be seen in the skull going from gibbons to chimpanzees to man?

The Fossil Record

If evolution has indeed occurred, we should expect to see evidence for it in fossils. The fossil record between apes and man is fairly good, with many intermediate forms preserved. Likewise, the fossil record tracing the descent of horses is good (Figure 13-5).

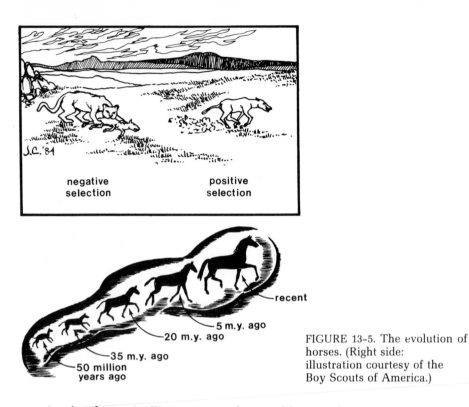

negative
selection

positive
selection

recent
5 m.y. ago
20 m.y. ago
35 m.y. ago
50 million
years ago

FIGURE 13-5. The evolution of
horses. (Right side:
illustration courtesy of the
Boy Scouts of America.)

8. As shown in Figure 13–5, why would natural selection favor large horses over small ones?

The fossil record is not as good for all species as it is for horses. The first multicelled organisms on earth, for example, appear "suddenly" in rocks 600 million years ago. It is possible that these multicelled organisms were the first to have hard parts that could be preserved in rock.

9. Why would an organism, such as a clam, be much more likely to become a fossil than a jellyfish?

Summary

Evidence for evolution can be seen in the field of dog breeding, where the many breeds of dogs we see today have been bred from a single ancestral species of wild dog. Charles Darwin first proposed the theory of evolution and put forth the idea of natural selection. According to natural selection, a changing environment selects certain individuals over others, and these individuals tend to pass on their traits to their offspring. Evidence for evolution can be seen by comparing bone structures between closely related living species and by examining the fossil record.

10. A knowledge of how evolution works is important to many people today. Animal breeders of all sorts breed those individuals that have certain desired traits. Plant breeders breed those plants that give the highest yield or those flowers that are the most unusual or pretty.

Evolution can be seen in the field of insect control, where the makers of pesticides have often been dismayed to find that a particular pesticide is effective for only a limited time. When a pesticide is sprayed on a population of mosquitos, for example, most of the mosquitos are killed. There are a few individuals, however, that are not susceptible to the pesticide. How can these few individuals cause the evolution of a mosquito that is immune to that particular pesticide?

Activities

- In a series of drawings show why some organisms in the same environment may have evolved so that they look alike. For example, the whale (a mammal) and the shark (a fish).
- How did the giraffe get its long neck?
- The human appendix and wisdom teeth are both examples of vestigial structures. Write a report on what vestigial structures are.
- It is said that the fossil record is like a history book of the earth with many of the pages missing. Explain.

14 ECOLOGY

Ecology is the study of how living organisms interact with their environment. It examines what is known as the "web of life."

Notice in Figure 14-1 that the web of life consists of plants, animals, and decomposers. Plants provide food for animals directly and indirectly. Some animals, such as insects, rabbits, and deer, feed on plants directly. Other animals, such as the fox and snake, feed on animals that feed on plants.

When plants and animals die they are decomposed by organisms such as bacteria and fungi. In this way they are returned to the soil.

1. The bird shown in Figure 14-1 feeds on plants both directly and indirectly. Explain.

2. Plants need both soil and water to grow. They also need one other key ingredient. What is it?

The web of life differs in appearance from one area to another. Northern areas, for example, have caribou instead of deer feeding on plant life. The plant life is also different because of the climate. The basic relationships between plants and animals remains the same, however.

3. Algae are one of the major food producers in the sea. These microscopic plants are the food source for many species of fish and for mammals, such as whales. The animals that feed on algae are equivalent to what four animals in Figure 14-1?

4. The shark is an animal that feeds upon smaller fish but is not itself preyed upon. The shark is equivalent to what animal in Fig. 14-1?

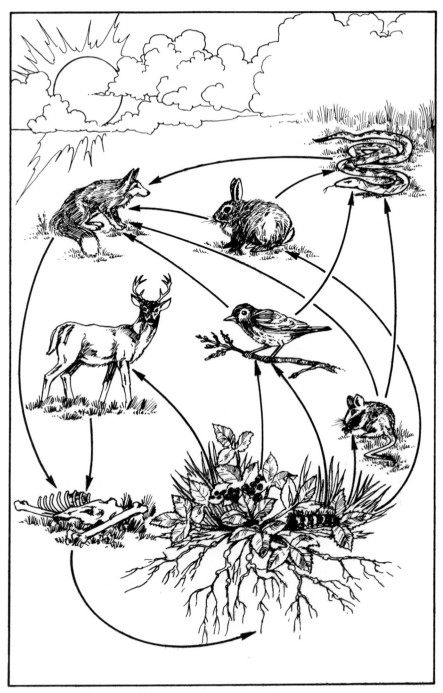

FIGURE 14–1. The web of life.
(Illustration courtesy of the
Boy Scouts of America.)

Habitat and Niche

Where an organism lives is referred to as its *habitat*. Examples of habitats include forests, grasslands, deserts, and oceans. The role an organism plays in its habitat is referred to as its *niche*. The niche of the rabbit in Figure 14–1, for example, is to feed on plants and in turn be fed upon by foxes and snakes. The niche of the deer is simply to feed on plants.

5. What is the niche of the plants in Figure 14–1?

6. What is the niche of bacteria and fungi in the soil?

Interrelationships between organisms in a habitat are often quite complex and fascinating. For one thing, there are often a great deal more niches than one would initially suspect. The rabbit shown in Figure 14–1, for example, can be subject to any of a number of parasites that live in or on its body (Figure 14–2). These parasites rely solely on the rabbit for their food supply.

7. The parasites shown in Figure 14–2 act to limit the rabbit population. How?

8. In some habitats organisms help each other in the battle for survival. The rhinoceros, for example, has parasites living on its back that serve

FIGURE 14–2. Some of the parasites that attack a rabbit.

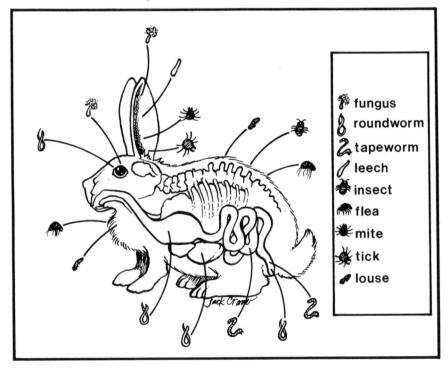

as food for a certain type of bird. This bird benefits from the food supply and in turn warns the rhino of any impending danger from predators. Other than signaling alarm, how else does this bird aid the rhino?

9. Many organisms change their role in the habitat from one season to another. A large number of plants, for example, only provide food in the summer months. This means that animals that depend on these plants for food must make provisions for the winter months. Insects do this by dying off in the cold weather and laying their eggs for the next spring. Birds can adjust by migrating to warmer areas where the food supply is more plentiful. Animals such as deer simply make do on the decreased vegetation that is available. How do bears adjust to the reduced food supply in winter?

Succession

Over a period of time, many habitats change with respect to the types of plants and animals that live there. This change is known as succession. Figure 14–3 shows how succession would occur on an abandoned field in a climate such as is experienced in the southeastern United States.

Succession occurs because plants and animals cause a change in the environment in which they live. The first weeds and grasses that appear on a bare field, for example, change the environment by shielding the soil from direct sunlight. As these plants spread, the ground surface becomes cooler and more moist than it was originally. Thus, the environment at the ground surface has been changed. The new surface conditions favor the sprouting of shrubs. As shrubs grow they shade out the grasses and also build up the soil in the area. In addition, they attract animals that also enhance the soil. Pine seedlings soon take hold and as they grow, they in turn shade out the shrubs. They are not able to shade out oak and hickory seedlings, however, that have found the forest floor suitable. These seedlings grow into large trees that eventually shade out the pines.

FIGURE 14–3. Succession on an abandoned field in the southeastern United States. (Dorfman, 1975.)

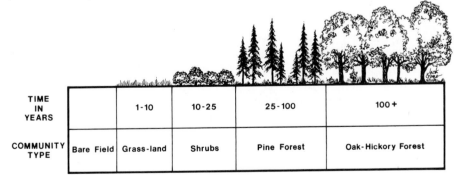

TIME IN YEARS		1-10	10-25	25-100	100+
COMMUNITY TYPE	Bare Field	Grass-land	Shrubs	Pine Forest	Oak-Hickory Forest

10. Fire is common to many pine forests in the southeastern United States. Fire periodically sweeps through the underbrush but does not affect the resistant pine trees. How does fire prevent an oak-hickory forest from developing in many areas?

11. Many shallow areas of the sea floor that are covered with sea grasses remain covered with this vegetation. That is, succession does not occur. Although the grasses shade out the sun from the sea floor, they do not appreciably change the temperature of the ocean bottom. Why not?

The Human Factor

No discussion of ecology would be complete without discussing the role of human beings in the environment. We as a species can prey on all other animals but are preyed upon by only very few. Gone are the days when lions and grizzly bears claimed a significant toll on human life. The few organisms that do prey on us are the disease-causing parasites. Many of these have been controlled by modern medicine. Thus, we occupy a unique place in the web of life. We are not only the top predator, we can also change or destroy the entire habitat.

12. The use of the gun has led to the decline of many species and the extinction of others. The passenger pigeon, for example, was a bird that once numbered in the billions. Today that bird is extinct due to overhunting. However, the gun is not the only way that animals can be destroyed. The bulldozer is equally effective. Yet most animals can flee from an oncoming bulldozer. Why then does this machine decrease their numbers?

13. As we come to better understand the web of life in an area, we should also develop a certain respect for it and treat it as if it were fragile, for in many cases it is. In Japan, for example, the English sparrow was deemed to be a nuisance and a hunting program was begun. The English sparrow's niche in the web of life was not considered, however, and as a result vast areas were soon plagued by insects. Why?

14. In the past we have been able to utilize the food web without destroying it. People have been fishing for centuries, for example, without depleting appreciably the number of fish in the sea. With modern fishing methods, however, great reductions in certain fish species have occurred. This is due in part to the fact that the fishing nets used are of a small enough mesh to catch all the fish in an area. What would be the long-term advantage of using a wider mesh net so that young fish could escape?

Summary

The interactions between organisms and their surrounding environment are known as the web of life. The web of life varies in appearance from one area to another, but its main concept of food producers, consumers, and decomposers remains the same. Where an organism lives is referred to as its habitat, and its role in the habitat is its niche. Many habitats change in appearance with time because organisms living there actually change the environment. This change is referred to as succession.

We human beings are the only species on earth capable of drastically changing a habitat. We have used this ability both wisely and unwisely in the past.

15. Salt marshes are habitats that are found in many coastal areas. They are characterized by swampy ground and slow-moving streams. These marshes are nutrient rich and, as a result, are teeming with plant and animal life. Many species of fish, for example, use these marshes as spawning grounds.

    Developers of resorts, condominiums, and other real estate are filling in salt marshes at an alarming rate. They do this to make prime waterfront land. Why does this practice of filling in salt marshes pose a severe threat to the fishing industry?

    ### Activities
    - Observe and report on a habitat that you are familiar with.
    - Write a "Rest in Peace" report on an animal or plant species that has become extinct.
    - List some reasons that a plant or animal species should not be allowed to become extinct.
    - Describe ways that human beings change a habitat from what it was earlier.
    - The concept of *niche* can be thought of in social terms. People you know occupy certain niches in society. What are these niches?

15

POPULATION

Figure 15–1 shows population growth versus time.

1. List on a piece of paper the years 1 A.D., 200 A.D., 400 A.D., 600 A.D., etc., up to the year 1800. Also list the years 1850, 1900, 1930, 1960, 1976, 1990, and 2000. Then from Figure 15–1 give the world's population at each of these dates.

As you can see in Figure 15–1, the world's population has increased dramatically in recent years. The population explosion of recent years is due mainly to a decrease in the death rate, rather than an increase in the birth rate. The death rate has declined mainly because of improvements in medicine.

2. Why do you think population growth in recent years has been referred to as the population "explosion"?

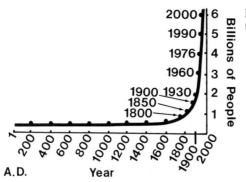

FIGURE 15–1. Population growth versus time. (Miller, 1979.)

Population Growth

To illustrate how population grows, let's look at two countries with quite different population growth rates. The two countries are Mexico and the United States.

TABLE 15-1 Population Growth in Mexico and the United States.

MEXICO: 3.4% GROWTH RATE		UNITED STATES: 0.6% GROWTH RATE	
Year	Population	Year	Population
1977	67 million	1977	218 million
1978	69.3	1978	219
1979	71.6	1979	221
1980	74.1	1980	222
1981	76.6	1981	223
1982	79.2	1982	225
1983	81.9	1983	226
1984	84.7	1984	227
1985	87.5	1985	229
1986	90.5	1986	230
1987	93.6	1987	231
1988	96.8	1988	233
1989	100.1	1989	234
1990	103.5	1990	236
1991	107.0	1991	237
1992	110.6	1992	238
1993	114.4	1993	240
1994	118.3	1994	241
1995	122.3	1995	243
1996	126.5	1996	244
1997	130.8	1997	246
1998	135.2	1998	247

Mexico has a growth rate of 3.4 percent, which means that there will be 3.4 percent more people living in that country next year than there are this year. In contrast, the United States has a growth rate of only 0.6 percent.

To find Mexico's population for any given year, multiply its population the previous year by 1.034 (its growth rate of 3.4 percent). To find the population of the United States for any given year, multiply its population the previous year by 1.006.

3. If Mexico has a population of 135.2 million in 1998, what will be its population in the year 2000?

In Table 15-1 notice that in the year 1998, Mexico's population will be double what it was in 1977. This means that the *doubling rate* for Mexico's population is 21 years. This doubling rate is among the highest in the world. In contrast, the United States has a doubling rate of 116 years.

4. If the present growth rate for the United States continues, in what year will this country have double the population it had in 1977?

The world in 1984 had a population of 4.8 billion people. This population was growing at a rate of 2 percent a year. Table 15–2 shows what the population will be in future years if this 2 percent growth rate should remain the same.

5. In what year will the world have double the population it had in 1984? What is the world's doubling rate?

TABLE 15–2 Growth of the World's Population.

1984 – 4.8 billion	1996 – 6.1	2008 – 7.7	2020 – 9.8
1985 – 4.9	1997 – 6.2	2009 – 7.9	2021 – 10.0
1986 – 5.0	1998 – 6.3	2010 – 8.0	2022 – 10.2
1987 – 5.1	1999 – 6.5	2011 – 8.2	2023 – 10.4
1988 – 5.2	2000 – 6.6	2012 – 8.4	2024 – 10.6
1989 – 5.3	2001 – 6.7	2013 – 8.5	2025 – 10.8
1990 – 5.4	2002 – 6.8	2014 – 8.7	2026 – 11.0
1991 – 5.5	2003 – 7.0	2015 – 8.9	2027 – 11.2
1992 – 5.6	2004 – 7.1	2016 – 9.0	2028 – 11.5
1993 – 5.7	2005 – 7.3	2017 – 9.2	2029 – 11.7
1994 – 5.8	2006 – 7.4	2018 – 9.4	2030 – 11.9
1995 – 6.0	2007 – 7.6	2019 – 9.6	2031 – 12.2

Future Population Growth

As you can see in Table 15–2, at the present rate of growth, the world's population will more than double in your lifetime. But is this growth rate likely to remain the same? With a third to a half of the world's population currently malnourished or undernourished, can the earth sustain a population more than twice the size it has now? There are some people who say we can and some who say we cannot. Nobody knows for sure.

We have made strides in feeding the world's population. For example, scientists have come up with new strains of wheat, rice, and other crops that yield quite a bit more than the old strains did. In countries such as Mexico, however, the growth in population has already outstripped the effects of these better-yielding crops. The result has been simply a greater number of people than ever before living at or below the subsistence level.

The current growth rate of the world cannot continue indefinitely, because the world obviously cannot support an infinite number of people. Our numbers will be held in check in the future, perhaps by mass starvation or by war. A better alternative is for us to do it voluntarily, by bringing our growth rate down to zero. This is what is called Zero Population Growth (ZPG). When the world reaches ZPG, the birth rate will match the death rate and our population will no longer increase.

6. A few of the European countries have reached zero population growth. Sweden is one of these countries. This country had a population of 8.3 million in 1984. What will its population be in the year 2000 if it maintains ZPG?

Efforts toward curbing population growth have been made in several countries. In India and China, the world's two most populous countries, measures have been tried such as mass sterilization, free dispersal of contraceptives, and financial incentives such as tax breaks for having a small family. In each of these countries couples have been educated to plan for a small family, rather than let the laws of nature dictate the size family they will have. China has had somewhat more success than India in curbing its population, although neither country has yet reached a growth rate of zero.

7. In 1984, India had a population of 750 million. China's population was 1 billion. What percentage of the world's 4.8 billion people lived in each of these countries in 1984?

Effects of Population Growth

Increases in population strain not only the world's food resources, but natural resources as well. People need resources to build and heat their houses. They need resources in many aspects of their lives. The more people, the more resources that will be needed.

Population growth also has an effect on the amount of pollution produced. In general, the higher the population, the more pollution produced.

8. Some people feel that the world is already overcrowded and that we should actually reduce our numbers somewhat. What effect would diminishing population have on our demand for natural resources and on the amount of pollution produced?

Population Distribution

The world is sometimes divided into two categories, based on standard of living. These two categories are the *first world nations* and the *third world nations*.

First world nations are nations that are industrialized. These are nations such as the United States, Canada, the European countries, Russia, and Australia. They contain 25 percent of the world's population and they use 80 percent of the world's natural resources. They are high polluters. These countries typically have a low population growth rate.

Third world countries include India, China, much of Africa, and much of South America. These countries have 75 percent of the world's population but use only 20 percent of the world's natural resources. They have a quite different standard of living than do the first world countries. Malnutrition is common, illiteracy is high, and the population growth rate is very high. The high population growth rate means that these countries either have a population problem now or will have one in the future.

9. It is estimated that a person in an industrial country can use well over ten times as much energy and natural resources as a person in a third world country. Given this, in what way are small population increases in first world countries as serious a problem as large population increases in third world countries?

Summary

The population of the earth has grown enormously in recent years. The rate of growth has been much faster in third world nations, such as Mexico, than in first world nations, such as the United States. The future growth of the world's population will depend on factors such as improvements in agriculture and efforts towards curbing population growth. Population growth is directly related to resource use and to the amount of pollution produced.

The population of the world is unevenly distributed, with 25 percent of the people living in the industrialized first world nations and 75 percent living in the third world nations.

10. The United States has a standard of living many countries of the world are trying to attain. Yet if the rest of the world did reach the standard of living in the U.S., the drain on the world's natural resources would be exhaustive. This country, with 6 percent of the world's population, uses 40 percent of the world's natural resources. If 100 percent of the world was at the standard of living of the United States, what would be the percentage of natural resources used?

Activities

- Research and write a report on a particular famine, such as the Irish potato famine.
- Research and write a report on the Tragedy of the Commons.
- List some diseases of the past that have now been largely controlled by modern medicine.
- Nehru, the leader of India from 1947–1964, once said, "Population control cannot solve any of the world's problems, but none of the world's problems can be solved without population control." List problems that affect the world and what influence population has on them.
- This problem illustrates the concept of doubling rates: A man offers you a job for one month. On the first day he will pay you one cent for your work. On the second day he will pay you two cents, on the third day four cents and on the fourth day eight cents. Your salary will likewise double for each of the thirty days in that month. Will you work for this man?

16 MINERAL RESOURCES

Mineral resources provide the structural support for our present day civilization. We rely on these resources for many aspects of our lives. These resources can be classified into two groups—*metalliferous deposits,* from which metals such as iron, aluminum, and copper are extracted, and *nonmetallic deposits* such as building stone, rock salt, and clay.

Many deposits, especially metalliferous deposits, must be concentrated before they can be mined. This is because their average abundance in the earth's crust is so small that it would not be economical to mine them otherwise. Figure 16–1 shows how mineral resources can become concentrated.

1. When molten rock cools, it crystallizes into *igneous rock.* In the process of cooling, some of the early formed crystals may settle to the bottom of the liquid. Others that are lighter in weight may float to the top and form a type of scum. Which deposit in Figure 16–1 formed in this way?

2. Molten rock often forces its way into surrounding rock through pre-existing fractures. In so doing, it expands the fractures and forms what are called *dikes.* These dikes tend to "bake" the rock they intrude into. In so doing they may concentrate important minerals and thus form an ore deposit (Figure 16–1, Deposit B). The largest of the three dikes in Figure 16–1 has intruded into shale, limestone, and sandstone. Which of these three rocks was the most susceptible to baking? Which was the least susceptible?

3. Hot watery solutions are often present with molten rock. These hot solutions are responsible for much of the baking of surrounding rock. When these solutions reach an environment that is much cooler than

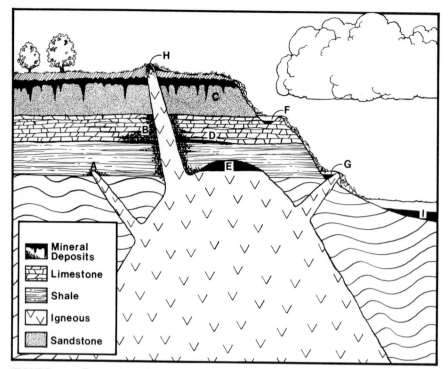

FIGURE 16–1. Formation of mineral resources.

where they originated, such as the ground surface, they deposit the minerals they had in solution. What deposit in Figure 16–1 formed in this way?

4. The boundary line between rock layers is often a zone of weakness along which igneous rock can intrude itself. What deposit in Figure 16–1 formed in this way?

5. Weathering of a deposit at the surface produces sediment. This sediment moves downslope under the force of gravity and eventually enters a stream. The stream may concentrate the deposit in certain locations such as pools or depressions on the stream bottom. These deposits are referred to as *placers*. What two deposits in Figure 16–1 are placer deposits?

6. Gold is often found as a placer deposit. A stream will sort gold from the other sediment because gold is much heavier. In areas where the current is weak, the gold will fall to the bottom but the remaining lighter sediment will continue to be carried. Gold has a specific gravity of 19 whereas quartz, a common sediment, has a specific gravity of only 2.65. About how many times heavier is a particle of gold than an equal-sized particle of quartz?

7. Deposits at Location A are not likely to be discovered. Even if they are, it is probably not economic to mine them. Why not?

8. Evaporation in a body of water such as a lake or enclosed bay can cause minerals such as gypsum and salt to precipitate out of solution and accumulate on the bottom. What deposit in Figure 16–1 formed in this way?

9. Some mineral deposits, such as iron ore, become enriched when downward moving ground water selectively dissolves that mineral. It then reprecipitates the iron at a lower level. Other deposits, such as aluminum ore, become enriched when ground water selectively dissolves and carries away parts of the surrounding material, leaving the ore behind. What deposit in Figure 16–1 was formed by the downward percolation of groundwater?

10. Unlike the deposits shown in Figure 16–1, rock deposits such as sand, gravel, and building stone do not need to be enriched before they become economical to extract. Yet most of the companies that extract these materials are local industries. That is, they supply only to a local area. Why would a builder in California probably not be interested in building stone from North Carolina, for example?

Methods of Extracting Minerals

Minerals are extracted from the ground by a number of ways. These include open pit mining, quarrying, underground mining, and dredging (Figure 16–2).

FIGURE 16–2. Methods of extracting minerals.

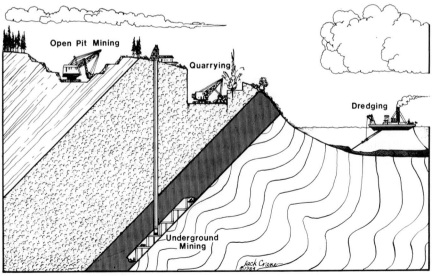

11. Open pit mining is generally cheaper and safer than underground mining. Why?

Processing

After a metalliferous deposit is mined, it must be processed. This means that the mineral must be separated from the rest of the rock. Methods of doing this are often complex, but many of them involve crushing the rock first. Processing methods have improved greatly in recent years so that ore that was once uneconomical to mine is being mined today.

12. In the 1920s, copper ores typically contained about 5 percent copper. Today, ores containing only 0.3 percent copper are mined. This is due in part to better processing techniques and in part to the fact that many of the high grade ores have been depleted. Processing lower grade ores produces more waste rock than was created previously. To get the same amount of copper from 0.3 percent ore, for example, a great deal more rock must be mined than with the 5 percent ore. How much more?

Conservation of Natural Resources

Some mineral resources are so abundant that it is unlikely that we will ever run out of them. Sand and gravel deposits, building stone, and aluminum ore, for example, are all quite abundant. Other mineral deposits, especially those containing metals, are more scarce. New discoveries will be found, but it should be kept in mind that there is a finite amount of these resources. They will not last forever.

To avoid running out of mineral resources in the future, we will have to alter our living habits so that we consume less of them. We will also have to practice conservation techniques such as *recycling*.

13. As some minerals become depleted, we can obtain more supplies by mining lower grade ore. This is the case with copper and iron, for example. Mining lower grade ore, however, generally takes more energy and is more destructive to the environment because of the large amount of waste produced. For some minerals, such as mercury, lower grade ores do not exist. Why can't ores of lower concentration be mined for these minerals?

14. Substitutions can be made for some resources that are growing scarce. The substitute is not always as good as the original, however. Aluminum can substitute for copper in wires, for example, but it does not conduct electricity as well as copper. New cars have considerably less metal than cars of the past. What has been substituted in place of parts that used to be made of metal?

Summary

Mineral resources provide the support for our present-day life. These resources form in a number of ways, as shown in Figure 16–1. Methods of extracting these resources include open-pit mining, quarrying, underground mining, and dredging. Once removed from the ground, these materials must be processed to separate the usable mineral from the waste material. Conservation techniques such as recycling will need to be practiced in the future to avoid running out of mineral resources.

15. Mineral resources are sometimes referred to as "one crop only" resources. What is meant by this?

   Activities
   - Pick a particular mineral resource. Research how it forms, how it is extracted and processed, and how it is used.
   - Read up on how various mineral deposits are processed.
   - Think of ways reusable bottles can conserve resources.
   - Obtaining minerals from the sea is not economically feasible for most minerals. Research why.
   - It is sometimes said, "You can't find gold in a silver mine." Research why this statement is not true.

17
ENERGY RESOURCES

It has often been said that the three basic essentials for life are food, clothing, and shelter. Today we could add energy as a fourth essential. We use energy to heat our homes, to run our appliances, and to power our automobiles. Energy is consumed in making the products we use and in transporting these products to us. Energy is also used in making fertilizers for crops and in running the machinery to tend and harvest these crops. Energy is very much a part of our daily lives. Without it we would have a much lower standard of living.

This exercise will acquaint you with energy resources that are being used today and with energy resources that may come into use in the future.

Oil and Natural Gas

Oil and natural gas are believed to have formed in continental shelf areas from the remains of aquatic plants and animals. These remains drift to the ocean bottom and accumulate with sediment that was brought there from rivers and waves attacking the shore. As this mixture of sediment and organic material becomes compacted under layers deposited on top of it, chemical reactions occur that over millions of years turn the organic matter into oil and gas.

Newly formed oil and gas tend to separate from the water that is also present. These hydrocarbons float upward through the sediment and may eventually escape at the surface. If an impervious rock layer is present, however, the oil and gas may accumulate to form a reservoir (Figure 17-1).

1. In Figure 17-1, well A is an exploratory well that shows only a trace of oil. Well B yields oil and well C yields natural gas. When pumping from well B begins, the oil level will eventually fall below the bottom

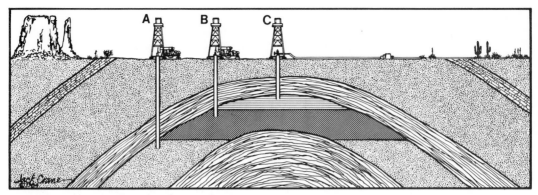
FIGURE 17-1. A trap for oil and natural gas.

of the well and no oil will be extractable. Pumping water into well A would allow well B to once again pump oil. Why?

Oil and natural gas account for 65 percent of the energy used in the world today. Although we are living in what has been called the "petroleum age," there are problems associated with the use of these fuels. For one thing, supplies are dwindling. The search for oil and gas has led to deeper wells than ever before and to areas of the world that are remote and inhospitable.

Another problem with these energy sources is that they pollute. Oil pollutes a good deal more than natural gas, but both of these fuel sources release carbon dioxide (among other pollutants) into the atmosphere. Carbon dioxide gas can increase the greenhouse effect of the earth's atmosphere and lead to an overall warming of the earth. This could change climates, which in turn would greatly affect farming areas. In a worst-case scenario it could melt the polar ice caps. This would raise sea level and flood many coastal areas.

2. Natural gas is used in cooking, home heating, and the generation of electricity. Name three ways that oil is used.

Coal

Coal forms in swamp areas such as shown in Figure 17-2. As generation after generation of plants live and die in these areas, thick deposits of peat build up. With time, these bottom peat layers compact under the weight of overlying material. After millions of years of heat and pressure, these layers gradually turn into coal.

3. Coal is mined both at the surface and underground. At the surface, strip mining is done by giant shovels that scoop out the coal. Underground mining is done by drilling "rooms" into a coal seam. When mining a room in a coal seam, some of the coal cannot be mined. Instead it must be left standing as pillars. Why?

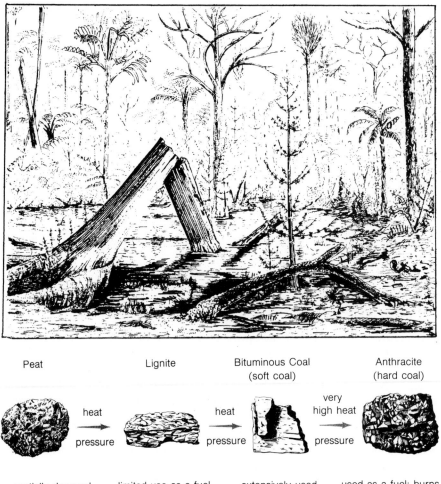

| Peat | Lignite | Bituminous Coal (soft coal) | Anthracite (hard coal) |

heat pressure → heat pressure → very high heat pressure →

partially decayed plant and animal matter in swamps and bogs; not a true coal

limited use as a fuel

extensively used as a fuel; burns with a smoky flame

used as a fuel; burns with a clean flame

FIGURE 17–2. The formation of coal. (Top: Bureau of Topographic and Geologic Survey, Pennsylvania Department of Environmental Resources; bottom: Miller, 1979.)

Coal accounts for about 30 percent of the energy used in the world today. It was used even more extensively in the past, however, before oil and gas came into widespread use. As petroleum supplies dwindle, the use of coal may once again climb.

Coal is not without its problems. Strip mining can permanently scar a landscape while water flowing through an underground mine can become polluted and enter a stream a bright orange color. This polluted water is known as acid mine drainage. In addition to these problems, the burning of coal causes air pollution. Acid rain is just one of the air pollution hazards associated with the burning of this fuel.

4.   The supply of coal left on the earth is about six times that of oil, or enough to supply the earth for 200 years. Some of this coal could be converted to oil by a process called *gasification*. This process produces only one-tenth the pollution that would be produced by burning the coal directly. Yet 30–40 percent of the energy content of the coal is lost in the process. Cite one advantage and one disadvantage of coal gasification.

Hydroelectric Power

Hydroelectric power is utilized in areas where a dam has been built on a river. Water released near the base of the dam spins turbines, which in turn generate electricity.

There are not many problems associated with hydroelectric power. The source of energy is relatively pollution-free, although habitats upstream are destroyed when the reservoir behind the dam first forms. Some dams also prevent the migration of fish, such as salmon, although this problem can be alleviated by building a stepped spillway to one side of the dam. The life of hydroelectric dams is limited to about 50–100 years because over time silt builds up behind the dam.

About 2 percent of the world's energy is generated by hydroelectric plants. This percentage is not likely to rise a great deal because dams have already been built on many of the favorable river sites. However, renovation could be done on many small hydroelectric plants that were abandoned in the past. Bringing these small plants into service once again would save utilities the cost of having to build new power plants.

5.   Towns located downstream from a hydroelectric dam might be concerned in times of flood. If a dam is properly designed and built, however, floods should not weaken it. After what other natural disaster might downstream towns want to evacuate?

Nuclear Power

Roughly 2 percent of the world's energy needs today are supplied by nuclear power. To generate this power, uranium atoms are bombarded by neutrons and thereby split in half. Tremendous amounts of energy are released in the splitting process. Indeed, one pound of enriched uranium can produce as much energy as twenty-five railroad cars of coal!

Producing uranium fuel is a relatively inexpensive process. Converting this fuel into electricity is not. Nuclear plants require advanced technology, which translates into high prices. In addition, many of the plants now operating have been plagued by problems, and as a result the building of nuclear plants has dropped off sharply.

There are problems with nuclear power other than the price. One of these problems is the question of where to dispose of the radioactive wastes generated by these plants. These wastes will stay radioactive for thousands of years and must therefore be buried in safe locations. Another problem is with the possibility of sabotage at a nuclear facility.

6. What was the first use of nuclear energy?

Other Energy Sources

Many energy sources that are not being used extensively today may be more widely used in the future. Among these sources are *geothermal energy, nuclear fusion,* and *solar power.*

Geothermal energy is energy from the earth's interior. The amount of energy present is enormous but in only a few areas of the world can it be profitably used. This is where the heat lies close to the earth's surface, such as in volcanic areas.

Nuclear fusion differs from nuclear fission in that it involves fusing atoms together instead of splitting them apart. When a fusion reaction occurs, tremendous amounts of energy are released. Creating this reaction is no simple process, however, and to date has not been achieved on more than an experimental scale. If it ever does succeed on a large scale, this could be a very important energy source.

FIGURE 17–3. Sail-assisted tanker, *Shin Aitoku Maru.* (Courtesy of Greg Davis.)

Solar energy is an energy source that will likely play a part in our future. It is currently used in some areas for hot water heating and home heating. In the future it will probably be used on a larger scale for generating electricity as well. Solar power and wind power (which is indirectly a form of solar power) are energy sources that are essentially nonpolluting. The only degrading of the environment that occurs is due to mining the raw materials for the manufacture of the solar collectors and windmills.

7. The ship shown in Figure 17–3 uses only about half the fuel oil as other ships its size. Why?

Conservation

Conservation is not often thought of as an energy resource but actually it is the least expensive and least polluting resource we have. If more people practiced conservation, our other energy resources would last longer and there would be less pollution.

FIGURE 17–4. (Illustration courtesy of the Boy Scouts of America.)

Family Tips for Saving Energy

8. Name ten ways that you could practice conservation where you live.

9. Conservation also means making changes in your lifestyle to conserve energy. This means driving smaller, more fuel-efficient cars or using mass transportation. How would the simple lifestyle change of living closer to work enable a person to save energy?

Summary

Energy is one of the basic essentials we need for life. Without it we would have a much lower standard of living. Major sources of energy currently being used include oil and natural gas, coal, hydroelectric power, and nuclear power. Other energy sources that may come into widespread use in the future include geothermal energy, nuclear fusion, and solar power. An important source of energy that is not often thought of as such is conservation.

10. Hydroelectric power and solar energy are considered renewable resources. That is, they can be used over and over again. Oil, natural gas, and coal are considered nonrenewable resources. What does this mean?

Activities

· What are some energy sources *not* mentioned in this exercise.

· Devise an energy plan for the future.

· It is said that coal may be the fuel of transition to the future. Do you know why? Create a "time line" diagram to explain how.

· One conservationist has said, "We believe in living off our income—the sun and the wind—and not off our principal—our deposits of fossil fuels." What did this person mean by this?

18 WASTE DISPOSAL AND POLLUTION

Our society generates wastes of all sorts. These wastes are in the form of solids, liquids, and gases. If the wastes are not properly disposed of (and often they are not), the result is pollution. In this exercise you will become familiar with the wastes our society generates and the ways we dispose of these wastes.

Solid Waste

Solid waste is the waste generated by homes and businesses. Table 18-1 shows a breakdown of this waste.

TABLE 18-1   Components of Solid Waste.

50%	paper
12%	glass
10%	metal
10%	food wastes
5%	plastic and rubber
3%	wood, dirt, grass, and leaves

In the past, solid waste was disposed of in town dumps. These dumps were usually unsanitary. Rodents thrived, fires were common, and the smell was often objectionable.

Most solid waste today is disposed of in *landfills*. Figure 18-1 shows a cross section of how an ideal landfill should be constructed. Although some

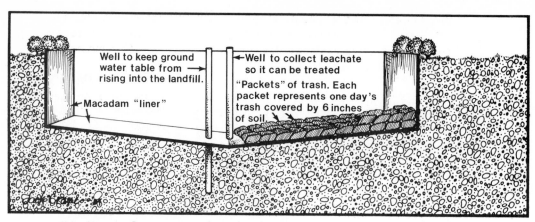

FIGURE 18–1. Cross section of an ideal landfill.

landfills have problems connected with them, most landfills are much preferable to town dumps.

One problem with landfills is the *leachate* they generate. Leachate forms when rainwater percolates down through trash and picks up impurities. Leachate is commonly full of iron oxide—rust—which gives it a bright orange color. If left untreated it will pollute a stream.

1. When some landfills reach their capacity, a plastic cover is put over them. A layer of dirt a few inches thick is then put over this plastic cover. How could a plastic cover drastically reduce the amount of leachate coming from a landfill?

2. Many towns and cities are facing the problem of their landfills filling up too rapidly. To remedy this situation, some municipalities have recommended recycling much of their solid waste. Looking at Table 18–1, list those waste products that you think could be recycled. Beside each waste product list its percentage of the total amount of solid waste.

Sewage

Sewage refers to the liquid wastes that households generate. In rural areas sewage is disposed of in septic tanks or cesspools. In municipal areas it is collected and transported in sewers. These sewers converge to one central location where the sewage is treated before releasing it into a nearby river or stream.

Sewage can receive three types of treatment—primary, secondary, or tertiary treatment. Primary treatment involves the removal of solids by the use of screens and settling tanks. This type of treatment removes about 50 percent of the impurities from sewage.

114

Secondary treatment removes approximately 90 percent of the impurities from sewage. This is done by first applying primary treatment to the waste, then using bacteria to break down remaining wastes. This method is now employed by most municipalities.

Tertiary treatment involves the further treatment of wastes so that the outflowing water is approximately 98 percent pure. This method of sewage treatment is expensive, however, and is not widely used.

3. Some cities have two separate sewer systems, one for sewage itself and another for precipitation runoff. Other municipalities have a combined sewer in which precipitation runoff travels with sewage in the same sewer. These latter systems sometimes experience severe problems with overloading at the treatment plant. That is, the treatment plant cannot handle all of the effluent at one time and must allow some untreated sewage to flow directly into the stream. This does not happen every day, however. When is it most likely to occur?

4. If you lived on the lower stretches of a major river, and your municipality used that river as its source of drinking water, why might you choose to buy bottled water?

5. Some coastal cities transport their sewage sludge out to sea by barge. This sludge is then dumped. List one danger that might result if this sewage is dumped too close to shore.

Waste Gases

Waste gases, when not properly disposed of, cause air pollution, a problem felt mainly in cities. Sources of air pollution include factories, automobiles, homes, and power plants. Because of the wide variety of pollutants that enter the air, the problems of air pollution can be quite complicated and not easily solved. In this exercise we will look at smog and at *acid rain*, a by-product of air pollution.

Smog. Smog forms when air pollutants enter the atmosphere and react with each other. The sun greatly aids these reactions, making smog worse on a sunny day.

Smog can be seen over a city on a day when there is little or no breeze to blow the pollutants away. If a stagnant air mass stays over a city for several days, as is often the case in Los Angeles, for example, smog can become quite a problem.

6. Why do mass transit systems, such as trains and buses, tend to reduce air pollution in a city from what it would be if we didn't have these systems?

Acid Rain. Acid rain forms as a result of the burning of coal and oil. Coal and oil contain small amounts of sulfur, which, when burned, reacts with oxygen and water vapor in the atmosphere to form a weak sulfuric acid. This diluted sulfuric acid then falls to the ground as precipitation and corrodes buildings, pollutes lakes, and disrupts ecosystems.

Acid rain can be controlled by installing scrubbers in smokestacks. These scrubbers remove sulfur before it can be released to the atmosphere. They are expensive to install and operate, however, and therefore many industries are reluctant to use them.

7. Canada receives acid rain that it claims originates in the United States. Yet if the smokestacks are in the United States, why should Canada experience the problem?

8. The world has much greater reserves of coal than of oil. This has led several people to suggest that power plants switch to burning coal instead of oil. But coal, as a rule, has more sulfur in it than does oil. Pertaining to acid rain, what would be the likely result of more power plants switching to coal?

Other Wastes

Other kinds of waste our society generates include mining wastes, chemical wastes, radioactive wastes, waste heat, and agricultural wastes. Some of these waste products can be considered hazardous, and thus their disposal is that much more difficult.

9. Water is used as a coolant for many industrial operations. If it is drawn from a stream for this purpose, it is usually returned to the stream in much the same chemical condition as it was originally. Yet the return water may readily kill fish and other organisms directly downstream. In what way has the returned water changed markedly?

Summary

We as a people generate many kinds of waste. The wastes we generate are in the form of solids, liquids, and gases. Some of the most common wastes are solid waste, sewage, and waste gases. Although these wastes are not always properly disposed of, the technology exists so that they can be.

10. One way to ease the problem of waste disposal is not to generate as many wastes in the first place. Name three ways that you yourself could reduce the amount of waste you generate.

Activities

- List local pollution problems in your community.
- Visit your local sewage treatment plant.
- Explain, by means of a diagram or chart, what is meant by the statement "dilution is the solution to pollution."
- Research and write a report on one type of waste and how it is properly disposed of.
- Show how fast-food places could reduce the amount of paper waste they generate.
- Read the book *Silent Spring* by Rachel Carson.

19
NATURAL DISASTERS

Earthquakes, volcanoes, landslides, floods, and hurricanes are just some of the natural disasters that strike the earth each year. We have little control in preventing these disasters, but we can do a lot to minimize the damage they incur. This exercise will acquaint you with natural disasters and how they can be dealt with.

Earthquakes

Earthquakes occur where two pieces of the earth's crust snap past each other (Figure 19–1).

Notice in Figure 19–1 that the two pieces of crust come under strain near the fault line. It is friction that keeps the two pieces from moving. As stresses continue to build up, however, this friction is finally overcome, and each piece snaps past the other causing the earthquake.

FIGURE 19–1. The formation of an earthquake. (A) Two crustal blocks are separated by a previous break in the rock; this break is called a fault line. (B) Strain develops along the fault line. (C) Strain overcomes friction causing an earthquake.

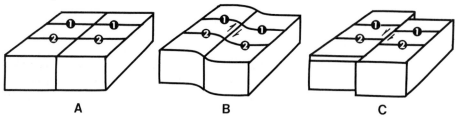

A   B   C

When an earthquake occurs, the movement is usually along a zone where the rock has been previously fractured. That is because as strain builds up it is more likely to release itself along a previous break than to actually cause new rock to break.

1. When picking a site for a new building, why is it best not to locate it near an old fault line?

2. Some scientists believe that major earthquakes can be prevented by inducing smaller ones instead. These smaller ones would be set off by pumping water into wells along the fault line. The water would lubricate the fault and hopefully cause small earthquakes to occur. A series of small earthquakes would relieve strain before it became great enough to cause a major earthquake. Why might you not want to experiment with this technique in a heavily populated area?

3. San Francisco was struck by an intense earthquake in 1906 (Figure 19–2). The earthquake toppled buildings and broke gas lines, causing fires that raged for days. Why do you think the city didn't use its water lines to put out these fires?

4. When an earthquake occurs on the sea floor, it creates a shock wave through the water. This shock wave may travel through the ocean for hours before reaching a coastal area. When it does reach the coast it may be as high as thirty to sixty feet (ten to twenty meters) and may cause considerable damage. Such waves are often referred to as tidal waves, but they actually have nothing to do with the tides. The correct term for such a wave is the Japanese term, *tsunami*. How might a coastal area know in advance that a tsunami might strike?

FIGURE 19–2. Earthquake in progress. (Fagan, 1965.)

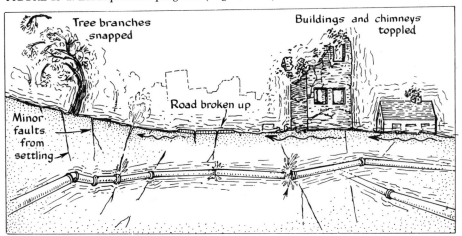

Volcanoes

Volcanoes are somewhat easier to predict than earthquakes. An increase in the amount of gases coming out of the top vent, for example, may foretell of an eruption. Today, geologists have installed sensitive tilt meters on the slopes of many volcanoes. These tilt meters record slight changes in the angle of the land, which may indicate when molten rock is moving beneath the surface.

Volcanoes have destroyed entire cities in the past. The city of Pompeii, Italy, for example, was buried by an eruption from Mount Vesuvius in 79 A.D. It stayed buried for nearly 1800 years before it was excavated by archeologists.

Volcanoes may erupt intermittently for years. Mount Saint Helens, for example, may continue to erupt for ten or fifteen more years before becoming dormant again.

5. Krakatoa was an island located northwest of Australia between Java and Sumatra. In 1883 much of this island was literally blown to bits by a volcanic eruption. The explosion was so great that it was heard in Australia, over 1200 miles (2000 kilometers) away! Dust-sized material from the eruption circled the globe for several years before settling back down to earth. This volcanic dust in the atmosphere caused the earth's temperature to be lower for several years afterward. Why?

Mass Movements

Mass movements, such as landslides and slumps, occur in areas where the land is steep. They occur because of slippage along a zone of weakness in the rock or soil (Figure 19–3).

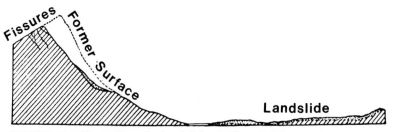

FIGURE 19-3. Formation of a landslide. (Miller, 1925.)

6. In a steep area, why might heavy rains cause more slumps and landslides to occur?

7. In Figure 19-4, houses A and B cost the same to build, but house B commanded a better view and therefore sold for more money. Why might house A be the better buy?

120

FIGURE 19–4. The locations of two houses.

Floods

Floods occur when an area receives a large amount of precipitation. The flooding is made worse if there is an accumulation of snow on the ground or if the water table is already high, thereby preventing much water from seeping into the ground.

The paving over of large areas of land has tended to make floods of today worse than in the past. Paving causes water to run off instead of seeping into the ground.

8. Why might flooding be worse in a city than in a rural area?

Figure 19–5 shows a floodplain and how it handles water during a flood.

FIGURE 19–5. Floodplain.

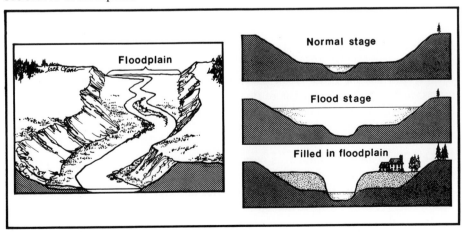

9. A hundred times more water may flow through a river during flood stage than at normal stage. Notice in Figure 19-5 the filled-in flood-plain. Why might people living on this fill area have a false sense of security?

Fire and Drought

Fire is a natural disaster that strikes large areas every year. It often works hand in hand with another natural disaster—drought.

10. Why are fires usually worse in times of drought?

11. A cigarette butt tossed out the window of a car will appear to "dance" across the highway. This is especially visible at night. How could this butt lead to a forest fire?

Hurricanes and Tornadoes

In the northern hemisphere, hurricanes form in tropical regions north of the equator. Once formed, they travel northward, bringing with them high winds and enormous amounts of precipitation.

12. Why are hurricanes that strike land more damaging than those that remain over the ocean for their entire life?

Tornadoes often form in times of severe thunderstorms. They are much smaller than hurricanes but they can be extremely destructive. They form when a funnel cloud descends to the surface of the earth. Wind speeds in a tornado commonly exceed 100 mph and sometimes reach 200 or 300 mph.

13. Why might more lives be lost if a tornado strikes at night than if it strikes during the day?

Insects and Disease

Insects and disease often work hand in hand in destroying large amounts of crops each year. In some areas of the world, insects have been known to literally run over entire villages. In other areas of the world, crops weakened by insects become more susceptible to disease.

14. The Panama Canal was completed in 1914. To complete this project, much brush had to be cleared and many swamps had to be drained. In the swamps lived an insect that transmitted both malaria and yellow fever. This biting insect can be found almost all over the world. What is it?

Summary

This exercise has acquainted you with natural disasters, how they occur, and how the damage from them can be minimized. For example, the damage from earthquakes can be minimized by simply not building in earthquake-prone areas. Likewise, the damage from landslides and slumps can be minimized by not building on steep slopes. Natural disasters cannot always be prevented, but the damage they cause often can be.

15. What is probably the easiest way to minimize the damage from flooding?

Activities

- A macabre game for stormy nights: Make a list of as many as possible of the potential disasters that could happen to this planet.
- Research a specific disaster that has happened in the past.
- List steps you would take to minimize damage from a specific natural disaster.
- Memorize steps to take in the event of an earthquake.
- Research modern architectural methods of safeguarding buildings from earthquakes.

20
LAND USE PLANNING

Land use planning is a practical application of environmental science. It involves planning the future land use of an area and it requires knowledge in several areas of science. It also requires an appreciation of such things as the culture of the people living in an area. In this exercise you will work with the same types of materials that planners use, so the evaluations and decisions you make will be similar to those made by planners.

The area to be studied is the town of Hockessin (Hō′-kes-in) in northern Delaware. This area has rich farmland and is characterized by rolling hills. It has a temperate climate, and the average rainfall is about forty inches per year.

1. In the late 1800s, the Hockessin area mined clay for making fine chinaware. This clay was mined in open pits. Many of these pits extended well below the water table, and as a result pumps were brought in to keep out the water. When clay mining became unprofitable around 1920, these clay pits turned into ponds. Why?

Road Map

A road map is indispensable to a planner for locating features in the area. This map is especially useful if field work is to be done in the area.

2. Two of the roads in this area have route numbers. What are the names of these roads and their route numbers?

3. Sewer lines have been laid in the Haverford and Gateway Farms developments (Figure 20–1, lower part). Sewers are needed whenever

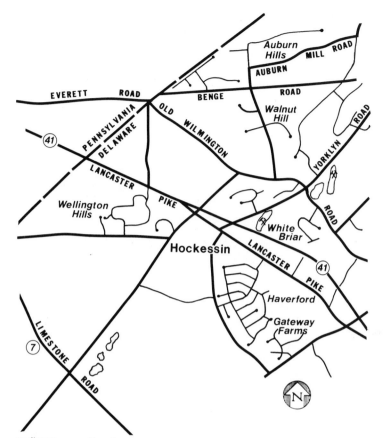

FIGURE 20-1. Road map.

there are a lot of houses in an area. Other housing developments in this area will need sewers in the future. Name three of these housing developments.

Large developments generally switch over to a central water supply, rather than each individual home having its own well. Before designing a water system for a development, however, it is first necessary to determine how much water that development is going to need. To determine this we use this formula:

Number of houses in development x 4 people per house x 100 gallons per person per day = water use

Notice that we use the average figures of 4 people per house and 100 gallons of water per person per day. These are good average figures.

4. How many gallons of water per day will the Haverford development require if there are 275 houses located in that development?

Geologic Map

Figure 20–2 is a geologic map of the area. A geologic map shows what type of rock is found at the surface of an area or, if soil covers the area, the first rock type that would be encountered if that soil cover were removed.

FIGURE 20–2. Geologic map.

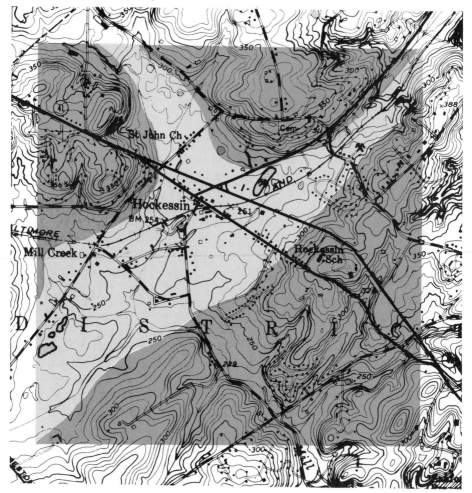

Cockeysville Marble. This formation is composed of *marble* that at one time was quarried for building stone. Ground water is found in the fractures of this rock. These fractures have grown larger due to ground water dissolving some of the marble. Thus, wells drilled into this formation may yield large quantities of water. An average well yield in this formation is 100 gallons per minute.

Wissahickon Formation. This formation is composed of a black shiny rock called *schist.* In this climate, schist is more resistant to weathering than is marble. It therefore underlies the uplands surrounding the marble valley. Ground water is also found in the fractures of this formation, but because water does not easily dissolve this rock, the fractures have remained small. Depending on the number of fractures intersected when drilling, a well in this formation will normally yield from 0–20 gallons per minute.

5. If you were to drill one well to provide for the water needs of the Haverford development, which formation would you choose to drill into and why?

6. Water use fluctuates in the course of a day. There are times of peak demand for water, such as at breakfast and supper, and times of little demand, such as in the middle of the night. A well might be able to handle a housing development's overall demand for water, yet be unable to handle its peak demand. Instead of drilling a second well, however, a storage tank could be built. How might a storage tank enable a well to handle the peak demand for water in an area?

Soils Map

A soils map shows what types of soil are in an area. These maps are useful to many individuals, particularly farmers and planners.

Soils Descriptions

Refer to Soils Map (Figure 20–3).

ChB2. Chester loam, 3–8 percent slopes, moderately eroded. This soil is medium textured, deep, and well drained. It is suitable for almost all uses. Being deep and well drained, it is satisfactory for septic systems.

GmB2. Glenelg and Manor loams, 3–8 percent slopes, moderately eroded. These are deep, well drained soils that are found on uplands. They are acceptable for septic systems.

GmC2. Glenelg and Manor loams, 8–15 percent slopes, moderately eroded. These soils are deep and well drained. They are found on uplands. If farmed,

FIGURE 20–3. Soils map. (USDA—Soil Conservation Service.)

measures should be taken to prevent erosion. These soils do not experience even a seasonal high water table. They are satisfactory for septic systems.

GnB2. Glenville silt loam, 3–8 percent slopes, moderately eroded. These soils are only moderately well drained. Because of this, drainage of the land is often needed before building. A high water table during parts of the year may cause septic systems to malfunction.

Ha. Hatboro silt loam, 0–3 percent. This is a deep, wet soil that is usually found in floodplains. It is very poorly drained and subject to severe flooding.

HbA. Hatboro silt loam, local alluvium, 0–3 percent slopes. This soil is found at the base of slopes, where soil material that washed from other areas has accumulated. It is poorly drained with the water table at or near the surface in winter and spring. Drainage of this soil for farming is possible if the drains are spaced close enough together. Much of the area covered by this soil is floodplain.

Soils maps are useful to builders in determining if an area is going to be suitable for septic systems. They are also useful in determining if a house is likely to have a wet basement or not.

7. From the description of soils given here, list by abbreviation which soils in this area are suitable for septic systems. Then list those that are only marginally suitable. Finally, list those soils that you think are unsuitable.

8. Notice in Figure 20-3 that the soils map is also an aerial photograph of the area. From this photograph, which would you estimate occupies more area—woods or farmland?

Future Development

Developing an area often means the paving over of areas that were formerly woodland or grassland. This developing has two potentially harmful effects. One is, precipitation falling on a paved area has little chance of seeping into the ground and replenishing the ground water supplies. Wells in the area must, therefore, curtail their pumping or they will run dry.

The second harmful effect of paving is the increased danger of flooding that results. Since water cannot seep into the ground, it runs off into the nearest stream. Flash flooding may result.

9. It is sometimes helpful to think in terms of extremes when trying to understand a problem. For example, let's suppose the entire Hockessin area is paved over. What would be the ground water future of this area if this should happen?

10. If this area were entirely paved over, what would happen after every heavy rainstorm?

11. In many areas of the country, floods today are worse than floods of the past, even though rainfall has not changed appreciably. Why might floods today be worse than floods of the past?

12. Sometimes dams are built to catch rain water runoff. This water is then used for a variety of purposes, one of which may be to recharge the ground water system. Some of these dams are designed to let the first ten minutes of runoff through before collecting the water. Why might we not want to use the first few minutes of runoff from an area?

13. If you were a planner and it was your job to locate a park in the area, what kind of an area would you look for?

14. There is no right or wrong place to locate a park. Some choices are better than others, however. Give two reasons why you might not want to locate your park next to the town's landfill.

Summary

In this exercise you have worked with the same types of materials that planners use. The road map is used for locating roads and housing developments in an area. The geologic map is used in deciding where to locate wells for ground water. The soils map is consulted before locating septic systems and for determining the type of vegetation in the locality. Based on these materials and an understanding of the culture of the area, plans about its future development can be made.

15. Residents in a locality are sometimes divided as to what their community's future should be. If a highway is to be built through the area, for example, some of the residents may be for it, and some may be against it. Give one reason why residents might like to see a highway built through their area. Give one reason why they might not.

Activities

- Collect other kinds of maps that planners might use, such as real estate maps, sewer maps, tax maps, etc. Put together a display of maps or a scrapbook of many kinds of maps.
- Write a short report on land use in your community.
- Visit a city or county planning office.
- Prepare a planning report for an area in your town.
- Explain why water use fluctuates in the course of a day and seasonally. Can you create a chart or diagram that shows the changes?
- Write a short report on what you would do with an area of land you are familiar with. For instance, if you owned five acres in the center of your town, what would you do with it? Sell the property? Build on it?

APPENDIX: ANSWERS TO CHAPTER QUESTIONS

Chapter 1
The Earth in Space

1. 100 billion.
2. Increases. (The more planets, the greater the likelihood of conditions favorable for life on at least one of them.)
3. Language would be one of the problems.
4. In our sun.
5. Jupiter, Saturn, Uranus, and Neptune.
6. The Asteroid belt.
7. It is much closer.
8. 3 A.M.
9. 24 hrs., 365 days.
10. More.
11. Cape Town is located in the southern hemisphere, which is experiencing winter.
12. Spring.
13. The sun is still shining at midnight during parts of the year.
14. 0.
15. B - 5° S latitude, 25° W longitude.
    C - 10° N latitude, 40° E longitude.

Chapter 2
The Interior of the Earth

1. 8,020 miles.
2. The weight of overlying rock becomes greater.
3. In the center of the earth.
4. Water.
5. Basalt.
6. They would first have to travel through the outer core, which they cannot do.
7. The crust is much thinner offshore.
8. S waves cannot travel through liquids, and the zone of partial melting has some properties of a liquid.
9. You would weigh more. The earth would be much denser, making its pull of gravity stronger.
10. We have no way of reaching there to measure it.

Chapter 3
Continental Drift

1. A fossil is an evidence of past life.
2. Mid-Atlantic Ridge.
3. Peru-Chile Trench.
4. Four.
5. Four.
6. Pacific Plate and North American Plate.
7. Part of the downturned crustal slab melted and rose to the surface to form these volcanic islands.
8. Indian Plate and Eurasian Plate.
9. Most of the earthquakes coincide with plate boundaries.
10. 5,000km × 100,000cm/km = 500,000,000cm.
    500,000,000cm ÷ 2.5cm/yr = 200,000,000 years.

Chapter 4
The Oceans

1. $\dfrac{4}{10,000} = \dfrac{6.5}{x}$ then cross multiply

    $4x = 65,000$

    $x = 16,250$ ft.

2. Ocean dumping of wastes, oil tanker spills, effluent from rivers, others.

3. Salt.

4. There would be so much, it would no longer be rare.

5. Peru Current, Falkland Current.

6. The Germans "cut" their engines and drifted through on the currents.

7. At low tide they could become stranded on land, without being able to go to sea. If at sea, they could not return to their port at low tide.

8. Shoreline B has a gradually sloping surface that helps to break the wave's force.

9. These cold, nutrient rich waters would support a great number of fish just as they do naturally off the coast of Peru.

10. If a smaller tanker has an accident, smaller amounts of oil are released to the ecosystem.

Chapter 5
The Hydrologic Cycle

1. Many outdoor plants have leaves in summer but not in winter.

2. Rainfall.

3. Dry.

4. Many answers possible.

5. Lower. Much water that had been in the sea was tied up as ice.

6. Sea level would rise and eventually flood these coastal areas.

7. It can evaporate or run off the land to a stream.

8. Ocean water is salt water.

9. Water in the oceans—97.2 percent, Ice—2.15 percent, Ground Water—.615 percent, Lakes and Inland Seas—.017 percent, Water in the Atmosphere—.001 percent, Streams—.0001 percent.

10. It is not salty.

Chapter 6
The Ground Water System

1. It would create a cone of depression (Figure 5–1).

2. Water enters an aquifer through rainfall (or snowmelt). Water can leave an aquifer by seepage into a stream (or the ocean, if it is nearby) or by pumping.

3. They came from small amounts of the rock that were dissolved in the water.

4. The stream loses water to the aquifer.

5. A—2 gpm; B—50 gpm; C—10 gpm.

6. The water table defines the top of the aquifer.

7. Water solution enlarges the fractures in limestone aquifers.

8. Movement of water, and hence pollution, can be much faster in a limestone aquifer because the water can travel quite rapidly through fractures.

9. Water shot to the surface. (The well became a gusher.)

10. They are both liquids.

Chapter 7
The Atmosphere

1. Dry Air: 78 percent nitrogen Humid Air: 76 percent nitrogen
   21 percent oxygen   20 percent oxygen
   1 percent other gases   1 percent other gases
                              3 percent water vapor

2.   13.65 g/cu meter at 100 percent humidity (saturation)
   − 4.85 g/cu meter at 36 percent humidity
   ───
      8.80 g/cu meter
   This cubic meter of air can hold 8.8 more grams of water.

3.   30.40 g/cu meter of water at 86° F and 100 percent humidity
   −17.31 g/cu meter of water at 68° F and 100 percent humidity
   ───
      13.09 g/cu meter
   This cubic meter of air will have given up 13.09g of water as rainfall.

4. 88° F − 32° F = 56°. It must cool.
   56° ÷ 3.3°/1000ft = 17,000 ft high.

5.   Equator: 1040 mph × 24 hrs = 24,960 miles
         60° N: 900 mph × 24 hrs = 21,600 miles
         30° N: 520 mph × 24 hrs = 12,480 miles
   North Pole: 0 mph × 24 hrs = 0 miles

6. Its relative humidity decreases because warmer air can hold more moisture.

7. This air is warming and will soak up any available moisture rather than give it up as precipitation.

8. It was spinning faster with the earth at the lower latitude (Figure 7–5) and thus "gains" on the earth at the higher latitude.

9. The warm moist air becomes chilled and therefore cannot hold as much moisture. It therefore gives up some in the form of precipitation.

10. Air pollution by the human population—the Industrial Revolution.

Chapter 8
Climates

1.

	Iquitos	Verkhoyansk
January	77° F	−60° F
February	77	−54
March	78	−28
April	79	− 2
May	76	28
June	75	51
July	75	59
August	76	54
September	77	40
October	77	12
November	78	−24
December	77	−51

2. Iquitos—average yearly precipitation—103" Verkhoyansk—average yearly precipitation—3.9" Iquitos receives so much rainfall because it is located in an area of rising air. Rising air cools and in so doing gives up moisture as precipitation. (Part of the reason why Verkhoyansk has so little precipitation is that it is located in an area of sinking air.)

3. St. Louis has an average temperature range of $77° − 34° = 43° F.$
San Francisco has an average temperature range of $62° − 48° = 14° F.$

4. New York is located on the other side of the continent from San Francisco. Westerly winds blow from New York City out toward the Atlantic Ocean, not vice-versa.

5. They are made worse by pollution from the city.

6. As air rises it cools and in so doing releases some of the moisture it was carrying.

7. July, August, September.

8. Its altitude is much higher.

9. It was cooler then than it is today.

10. The continent could drift closer to the equator.

Chapter 9
Landscapes I:
Weathering and Erosion

1. It seldom gets above freezing in polar areas.

2. Tree roots, frost wedging, burrowing by animals, abrasion by people.

3. Water is needed for chemical weathering.

4. Frost action causes fractures in the rock to become longer. Chemical weathering then has a greater area over which to work.

5. They may tilt downslope.

6. They would lubricate the zones of weakness along which slumps can form.

7. It could create steep slopes that would be susceptible to slump.

8. Glaciers carry the material in the ice.

9. Soil particles are no longer held together by moisture.

10. Yes. We have plenty of areas in the world covered by soils.

Chapter 10
Landscapes II:
The Influence of Climate

1. Rainfall is much greater in a rainforest.

2. This area receives an enormous amount of rainfall.

3. In a temperate landscape, frost action would be greatest in winter and nonexistent in summer.

4. The tree roots would expand more in the summer and hence expand any rock fractures they may have grown into.

5. It loses its current and thus its ability to transport sediment.

6. The U-shaped valley on the right.

7. It could carry the smaller material but not the larger material.

8. In an arid landscape, there is little or no vegetation to hold soil in place.

9. Intense wave action carries away sediment.

10. Tropical Landscapes—running water.
Temperate Landscapes—running water, gravity processes.
Polar Landscapes—ice.
Arid Landscapes—running water (also wind).
Coastal Landscapes—running water.

Chapter 11
Soils

1. Water runs off a steep area at a faster speed so it carries more sediment.

2. No. The soil did not develop in place so there is no reason to believe that the rock fragments in the C horizon would match the underlying bedrock.

3. Less distinct. These organisms tend to mix up the soil.

4. They are returned to the A horizon when the plant dies and decays.

5. The microorganisms that are very active in this climate decay dead plant matter before it can build up into humus.

6. Air.

7. Soil is being eroded 100 times faster than it is forming in such areas. Six inches of topsoil will be removed in 90 years.

8. Gullies.

9. Soil erosion adds to stream pollution.

10. Underneath the soil.

Chapter 12
History of the Earth

1. It is more dense than most other material.

2. Volcanoes.

3. Lightning.

4. Precambrian: 4 billion years
   Paleozoic: 375 million years
   Mesozoic: 165 million years
   Cenozoic: 65 million years

5. Paleozoic.

6. Paleozoic.

7. Precambrian.

8. Mesozoic.

9. $\dfrac{10,000,000 \text{ yrs.}}{4,600,000,000 \text{ yrs.}} = \dfrac{1}{460}$ the age of the earth.

10. Cephalopods, Rugose (horn) corals, Tabulate coral, Bryozoan, Trilobite. The giant Cephalopod is about to feed on a Trilobite.

11. They had not yet appeared.

12. It formed from the remains of dead organisms, that is, fossils.

13. Tyrannosaurus rex.

14. Horns (and a type of shield behind its head).

15. Active volcanoes are spewing gases and dust into the atmosphere.

16. Palm trees.

17. North America and Africa drifted apart.

18. North America was closer to the equator in late Paleozoic time.

19. North Dakota, South Dakota, Nebraska, Wyoming, Louisiana, Florida.

20. Nuclear war, pollution (to name two).

Chapter 13
Evolution

1. Dog breeders would have looked for individual dogs with small stature and a long snout. A dachshund is such a dog.
2. Farmers scatter its seeds.
3. Fishes.
4. Protozoa.
5. One way is that individuals can warn the entire group of impending danger.
6. By talking to a genetic counselor they could find out their chances of passing the defect along to future generations.
7. The skull gets larger.
8. They could run faster.
9. A clam has a hard shell.
10. The few individuals that are immune to the pesticide will breed with each other. Many of their offspring are likely to be immune to the pesticide as well.

Chapter 14
Ecology

1. It feeds directly on the berries of the plant and indirectly on insects that feed on the plant.
2. Sunlight.
3. Deer, birds, mice, rabbits (and insects also).
4. The fox.
5. The plants provide food for other organisms.
6. The bacteria and fungi are responsible for causing dead organisms to decay back into soil.
7. Some of the rabbits are killed by the parasites.
8. It eats parasites living on the rhino's back.
9. They hibernate.
10. It kills off the young seedlings.
11. Water changes temperature more slowly than land. It can also circulate.
12. The bulldozer destroys their habitat.
13. The English sparrow fed on insects.
14. The young could escape to breed a future generation.
15. Fish lay their eggs in these salt marshes. If they are filled in, there will be no future generations of these fish species.

Chapter 15
Population

1.

1 A.D.	0.3 billion	1800	0.9 billion
200 A.D.	0.3 billion	1850	1.1 billion
400 A.D.	0.3 billion	1900	1.6 billion
600 A.D.	0.3 billion	1930	2.0 billion
800 A.D.	0.3 billion	1960	3.0 billion
1000 A.D.	0.3 billion	1976	4.0 billion
1200 A.D.	0.4 billion	1990	5.0 billion
1400 A.D.	0.4 billion	2000	6.0 billion
1600 A.D.	0.5 billion		

2. The increase in numbers has been explosively rapid.

3. 135.2 million × 1.034 (growth rate) = 139.8 million (1999 population). 139.8 million × 1.034 = 144.6 million

4. 1977 + 116 years (doubling rate) = 2093 will be the year the U.S. will have double the population it had in 1977.

5. 4.8 billion × 2 = 9.6 billion. According to Table 15.2 the world will have 9.6 billion people in the year 2019.

6. 8.3 million.

7. India $\dfrac{750 \text{ million}}{4,800 \text{ million}} = \sim 16\%$

China $\dfrac{1 \text{ billion}}{14.8 \text{ billion}} = \sim 21\%$

8. Diminishing our numbers would lessen the demand for natural resources and decrease the amount of pollution produced.

9. Small population increases in first world countries mean large increases in the use of energy and natural resources.

10. $\dfrac{6\%}{40\%} = \dfrac{100\%}{x}$; cross multiplying: 6x = 4000; x = 666%

Chapter 16
Mineral Resources

1. E.

2. Most susceptible—limestone; Least susceptible—sandstone.

3. H.

4. D.

5. F and G.

6. About seven times heavier.

7. The deposit is quite small and is deeply buried.

8. I.

9. C.

10. It would cost too much to transport the stone from North Carolina to California.

11. Cheaper: Material does not have to be transported to the surface.
    Safer: No danger of cave-ins with open pit mines (or dust and fumes).

12. About seventeen times more waste.

13. They don't exist.

14. Plastic.

15. Supplies cannot be replenished as with farm crops.

Chapter 17
Energy Resources

1. Oil floats on top of water. Pumping water into well A would cause the oil above it to float to a higher level.

2. Heating oil, gasoline for cars and trucks, manufacture of polyester and other synthetic fibers, manufacture of plastics, as a lubricant, and other uses.

3. To prevent cave-ins.

4. Advantage: more oil could be obtained (and less pollution produced).
   Disadvantage: a significant amount (30–40%) of the energy content is lost.

5. An earthquake.

6. The atomic bomb.

7. This ship has two sails to help power it.

8. See Figure 17–4.

9. Living closer to work decreases the amount of energy needed to get back and forth.

10. Nonrenewable resources can be used only once. They cannot be replenished.

Chapter 18
Waste Disposal and Pollution

1. The plastic cover should prevent rain water from percolating through the landfill and becoming leachate.

2. Paper   50 percent
   Glass   12 percent
   Metal   10 percent

Note: Some of the remaining solid waste is recyclable also, but it is not generally practical to do so.

3. After heavy rains or intense snow melts.
4. Towns and cities upstream have probably dumped waste water into the river.
5. The sewage may wash back to the beaches.
6. When trains and buses are used, fewer cars are driven. (One bus carrying thirty people pollutes less than thirty individual cars.)
7. The wind blows it into Canada.
8. More acid rain would be produced.
9. It is much warmer than it was.
10. Many possible answers.

Chapter 19
Natural Disasters

1. If there are stresses in the area, they would be more likely to release along an old fault line than elsewhere.
2. If a major quake was induced by mistake, the results could be disastrous.
3. The water lines also broke.
4. They might be forewarned of a tsunami if a major earthquake had been reported.
5. The volcanic dust acted to shield out some of the sun's rays.
6. Heavy rains could lubricate zones of weakness in a steep area.
7. House B might be susceptible to slump.
8. Paving in the city will not let water soak into the ground as it will in rural areas. Thus, more water runs off.
9. Although they are elevated well above the river, flood waters will rise higher than before.
10. The area is much drier.
11. It could lead to a forest fire if it ends up in woods or brush along the roadside.
12. Civilization exists mainly on land.
13. During daylight hours, people may see it coming and be able to take shelter.
14. Mosquito.
15. Don't build on floodplain.

Chapter 20
Land Use Planning

1. Ground water that had previously been pumped out was allowed to fill up these pits.

2. Lancaster Pike—Rte. 41;
   Limestone Road—Rte. 7.

3. Any of the larger housing developments listed on the map.

4. 275 houses × 4 people per house × 100 gallons per person per day = 110,000 gallons.

5. The well should be drilled into the Cockeysville Marble because this formation is more likely to yield large quantities of water.

6. Water could be pumped into it during non-peak times and released during peak times when the well cannot keep up with demand.

7. Suitable: ChB2, GmB2, GmC2
   Marginal: GnB2
   Unsuitable: Ha, HbA

8. Farmland.

9. The ground water table would go down because ground water supplies could not be replenished by rain water.

10. Flash flooding would result.

11. More areas are paved today than in the past.

12. It contains much dust and debris that has settled to the ground.

13. Any reasonable answer is correct.

14. Noisy machines, blowing litter, objectionable smell.

15. For—could bring in commerce and industry.
    Against—could destroy a way of life they have become accustomed to.

BIBLIOGRAPHY

ADAMS, GEORGE F., and JEROME WYCKOFF. *Landforms.* Racine, Wisconsin: Western Publishing Co., Inc., 1971. This Golden Guide book discusses how landscapes form.

AHRENS, C. DONALD. *Meteorology Today.* St. Paul, Minn.: West Publishing Co., 1982. This introductory meteorology text has many good photographs and illustrations.

BATTAN, LOUIS J. *Fundamentals of Meteorology.* Englewood Cliffs, N.J.: Prentice-Hall, Inc., 1971. Various aspects of the atmosphere as well as climates are discussed.

BEISER, ARTHUR, and the EDITORS OF "LIFE." *The Earth.* Alexandria, Virginia: Time-Life Books, Inc., 1963. A well-illustrated book dealing with aspects of the earth such as its history, landscape processes, mineral resources, and energy resources.

BERMAN, LOUIS, and J.C. EVANS. *Exploring the Cosmos.* Boston, Mass.: Little, Brown and Company, 1980. A good book on astronomy.

BOLT, BRUCE A. *Earthquakes: a primer.* San Francisco, Calif.: W.H. Freeman and Co., 1978. A good short book on earthquakes and continental drift.

BOY SCOUTS OF AMERICA. *Environmental Science.* Irving, Texas: Boy Scouts of America, 1983. This merit badge booklet discusses many aspects of environmental science as well as careers in the field.

CARGO, DAVID N., and BOB F. MALLORY. *Man and His Geologic Environment.* Reading, Mass.: Addison-Wesley Publishing Co., 1974. This book covers a wide range of subjects including population, water resources, soils, mineral and energy resources, natural disasters, land use, and waste disposal.

CARSON, RACHEL. *Silent Spring.* Boston, Mass.: Houghton Mifflin Co., 1962. A "must" reading for environmentalists.

COMMONER, BARRY. *The Closing Circle.* New York, N.Y.: Alfred A. Knopf, Inc., 1971. A good book on the environmental crisis.

DORFMAN, LEON. *The Student Biologist Explores Ecology.* New York, N.Y.: R. Rosen Press, 1975. A good little book on the basic principles of ecology.

DUNBAR, CARL O. *Historical Geology.* New York, N.Y.: John Wiley & Sons, Inc., 1960. A good basic historical geology text.

EHRLICH, PAUL R., ANNE H. EHRLICH, and JOHN P. HOLDREN. *Ecoscience: Population, Resources, Environment.* San Francisco, Calif.: W.H. Freeman and Co., 1977. A very large and very comprehensive book dealing with the earth and its environmental crisis.

EHRLICH, PAUL R. *The Population Bomb.* New York, N.Y.: Random House, Inc., 1978. A description of the population explosion and what can be done about it. An excellent little book.

ELDREDGE, NILES. *The Monkey Business.* New York, N.Y.: Washington Square Press, 1982. An excellent short book on evolution.

FAGAN, JOHN J. *Modern Earth Science.* New York, N.Y.: CBS College Publishing, 1983. A well-written and well-illustrated textbook of geology.

FLINT, RICHARD FOSTER, and BRIAN J. SKINNER. *Physical Geology.* New York, N.Y.: John Wiley & Sons, 1977. A well-written and well-illustrated textbook on the science of geology.

FOX, WILLIAM. *At the Sea's Edge.* Englewood Cliffs, N.J.: Prentice-Hall, Inc., 1983. An excellent book for learning about the physical processes of the shore.

LEET, L. DON, SHELDON JUDSON, and MARVIN E. KAUFFMAN. *Physical Geology.* 6th ed. Englewood Cliffs, N.J.: Prentice-Hall, Inc., 1982. A good comprehensive textbook on geology.

LEHR, PAUL E., R. WILL BURNETT, and HERBERT S. ZIM. *Weather.* Racine, Wisconsin: Western Publishing Co., Inc., 1975. A Golden Guide book that discusses the atmosphere, weather, and climates.

MCALESTER, A. LEE. *The Earth.* Englewood Cliffs, N.J.: Prentice-Hall, Inc., 1973. An introductory textbook to geology and geophysics.

MARX, WESLEY. *The Oceans, Our Last Resource.* San Francisco, Calif.: Sierra Club Books, 1981. Describes our past use and misuse of the oceans and future prospects.

MATTHEWS, W.H., III. *Texas Fossils.* Austin, Texas: Bureau of Economic Geology, University of Texas, Austin. A book on the fossil record in Texas.

MILLER, G. TYLER, JR. *Living in the Environment.* Belmont, Calif.: Wadsworth Publishing Co., 1979. This book is comprehensive on the subjects of population, resources, pollution, and ecology.

MOORE, RAYMOND C. *Introduction to Historical Geology.* New York, N.Y.: McGraw-Hill Book Company, Inc., 1958. An introductory textbook to historical geology.

MOORE, RUTH, and the EDITORS OF TIME-LIFE BOOKS. *Evolution.* Alexandria, Virginia: Time-Life Books, Inc., 1970. Discusses how Darwin formulated the theory of evolution and how the theory operates.

NAMOWITZ, SAMUEL N., and DONALD B. STONE. *Earth Science: The World We Live In.* New York, N.Y.: D. Van Nostrand Co., Inc., 1960. A textbook in earth science.

PANANIDES, NICHOLAS A., and THOMAS ARNY. *Introductory Astronomy.* Reading, Mass.: Addison-Wesley Publishing Co., 1979. A good introductory book on astronomy.

PRESS, FRANK, and RAYMOND SIEVER. *Earth.* San Francisco, Calif.: W.H. Freeman and Co., 1974. A comprehensive textbook on geology.

RAMSEY, WILLIAM L., and RAYMOND A. BURCKLEY. *Modern Earth Science.* New York, N.Y.: Holt, Rinehart and Winston, Inc., 1961. A textbook on earth science.

RAYMO, CHET. *Biography of a Planet.* Englewood Cliffs, N.J.: Prentice-Hall, Inc., 1984. A good book tracing the physical history of the earth.

RAYMO, CHET. *The Crust of Our Earth.* Englewood Cliffs, N.J.: Prentice-Hall, Inc., 1983. This book covers subjects dealing with the earth's crust including continental drift.

SAGAN, CARL. *Cosmos.* New York, N.Y.: Random House, Inc., 1980. This book describes how science and civilization grew up together.

STOKES, WILLIAM LEE, and SHELDON JUDSON. *Introduction to Geology.* Englewood Cliffs, N.J.: Prentice-Hall, Inc., 1968. A well-illustrated introductory textbook on geology.

STRAHLER, ARTHUR N. *Introduction to Physical Geography.* New York, N.Y.: John Wiley & Sons, Inc., 1970. A well-illustrated book on physical geography.

TARLING, DON, and MAUREEN TARLING. *Continental Drift.* Garden City, N.Y.: Anchor Press/Doubleday, 1975. A short, concise book on the topic of continental drift.

UDALL, STEWART L. *The Quiet Crisis.* New York, N.Y.: Holt, Rinehart & Winston, Inc., 1963. This book traces the history of land use in America. A classic.

WEAVER, KENNETH F. *Energy.* Washington, D.C.: National Geographic Society, 1981. A well-written and well-illustrated report showing our current energy picture.

WOODBURN, JOHN H. *Energy.* Irving, Texas: Boy Scouts of America, 1982. This merit badge booklet discusses both energy use and energy conservation.

INDEX

p. 61 "temperate"   improper use of Mercator,
                              as on p. 63 — "arid areas of world"
good diag., many & useful even tho small & simple
some exercises too general, but OK to be optional
numerous questions, perhaps too many
good disc. of evolution
short ch. o k
legend omitted fm Fig 20-2 on p. 126
Ans. 5 on p. 134 (for Ch. 7) inverted 30 & 60° N
good idea to incl ans, in back